The Chemical Languages of the Nervous System

Josef Donnerer Fred Lembeck

THE CHEMICAL LANGUAGES OF THE NERVOUS SYSTEM

HISTORY OF SCIENTISTS AND SUBSTANCES

6 figures, 1 in color, 2006

KARGER

Basel · Freiburg · Paris · London · New York · Bangalore · Bangkok · Singapore · Tokyo · Sydney

Prof. Dr. Josef Donnerer
Institut für Experimentelle und
Klinische Pharmakologie
Medizinische Universität Graz
Universitätsplatz 4
A–8010 Graz (Austria)

Prof. Dr. Fred Lembeck
Institut für Experimentelle und
Klinische Pharmakologie
Medizinische Universität Graz
Universitätsplatz 4
A–8010 Graz (Austria)

Library of Congress Cataloging-in-Publication Data

Donnerer, Josef.
The chemical languages of the nervous system : history of scientists and substances / Josef Donnerer, Fred Lembeck.
p. ; cm.
Includes bibliographical references and index.
ISBN 3-8055-8004-5 (hard cover: alk. paper)
1. Loewi, Otto, 1873-1961. 2. Dale, Henry H. (Henry Hallett), Sir, b. 1875. 3. Feldberg, W. S. 4. Neurotransmitters–History. 5. Neuroscientists–Biography. 6. Biochemists–Biography. I. Lembeck, Fred. II. Title.
[DNLM: 1. Loewi, Otto, 1873-1961. 2. Dale, Henry H. (Henry Hallett), Sir, b. 1875. 3. Feldberg, W. S. 4.
Neurotransmitters–history.
5. Neurochemistry–Biography. QV 11.1 D686c 2006]
QP364.7.D65 2006
573.8–dc22
2005033505

P.O. Box, CH–4009 Basel (Switzerland)
Printed in Switzerland on acid-free paper by Reinhardt Druck, Basel
ISBN 3–8055–8004–5

Wilhelm Feldberg, Sir Henry H. Dale and Sir John H. Gaddum

In grateful memory to my teacher
in pharmacology
Sir John Henry Gaddum

F.L.

Printed with support of the
Bundesministerium für Bildung, Wissenschaft und Kultur, Vienna
OELZELT-Stiftung der Österreichischen Akademie der Wissenschaften, Vienna
Land Steiermark
Stadt Graz
Mundipharma GmbH, Vienna
Fresenius Kabi Austria GmbH, Graz
Dr. Molassa + Merz GmbH, Vienna
Gebro Pharma GmbH, Fieberbrunn
Novartis Pharma GmbH, Vienna
ZAK-Pharma Dienstleistungs GmbH, Vienna

CONTENTS

PREFACE

Documentation Otto Loewi

The collection of documents of Otto Loewi and his family consisted in 1962 primarily of a large suitcase full of unselected material that was in the hands of his son Guido. It was filed carefully by Wolfgang Giere and supplemented with material of various sources between 1962 and 1965. The collection of 13 bureau files was handed over by Guido Loewi to the Royal Society on November 17, 1965, and is stored there under the name 'Documentation Otto Loewi'. It served as the basis for the book by Lembeck and Giere [1968]: *Otto Loewi: Ein Lebensbild in Dokumenten.*

During the following 30 years, additional material came into my hands. Other documents resulted from a search of Anton Kernbauer in archives and elsewhere. The resulting *Supplement 1999,* collected in 3 bureau files, was added to the existing collection at the Royal Society on June 5, 2000.

Not included in the *Supplement 1999* are the transcripts of interviews of Ulrich and Anna Weiss in November 1990, done as part of the ICCGW's Holocaust Oral History Project, amounting to 130 typed pages.

The continuing support by Anna Weiss-Loewi and Guido Loewi has to be mentioned in particular. Numerous talks with Anna Weiss-Loewi and her husband Ulrich, and with Guido provided the information condensed in the chapter 'The Loewi Family's Way to the New World' (see pp. 11–29). Generally, I think it can be said that the life of Otto Loewi is better documented than that of any other prominent scientist.

Insert

When the recollections of only one contributor have been recorded, his initials are inserted in parentheses at the end of the sentence (e.g. F.L. = Fred Lembeck or J.D. = Josef Donnerer). As the older contributor, I (F.L.) would like to thank my friend and co-author J.D. We would both like to express our gratitude to the head of our department Prof. Dr. Bernhard A. Peskar for his continuous support.

Reference

Lembeck F, Giere W: Otto Loewi: Ein Lebensbild in Dokumenten. Berlin, Springer, 1968.

INTRODUCTION

The first localization of a receptor was achieved by Claude Bernard [1878] when he demonstrated by simple experiments that curare acts on the motor end plate without influencing nerve fibres or the central nervous system.

The English physiologist John Newport Langley (1852–1926) arrived at the concept of receptor sites for drug action from a study of pilocarpine and atropine on salivary secretion, which led him to conclude in 1878 that 'there is some "receptor" substance with which both atropine and pilocarpine are forming compounds according to some law in which their relative mass and chemical affinity for the substance are factors'. Langley [1905] concluded that 'there is a chemical combination between the drug and a constituent of the cell – the receptive substance'.

The German physician, biologist and chemist Paul Ehrlich was responsible for a more general concept of receptors, based on his work with tissue stains, immunity and chemotherapy. Ehrlich postulated (1885) that the specific activities of cellular protoplasm could be moderated by 'side chains' or 'receptors', to which a haptophoric group of drugs or toxins was attached and if the drug or toxin also contained a toxophoric group, the cell was damaged or killed. Ehrlich's emphasis on the binding of drugs to protoplasmic side chains (receptors) was summed up (1913): 'If the law is true in chemistry that corpora non agunt nisi liquida [drugs will not work unless they dissolved], then for chemotherapy the principle is true that corpora non agunt nisi fixata [drugs will not work unless they are bound]' [see Bowman and Rand, 1980; Rang et al., 2003].

Alfred Joseph Clark (1926) quantified the relationships of dose and its effect of agonists alone and also under the influence of an antagonist. He used acetylcholine as an agonist and atropine as its antagonist on a simple preparation, a strip of the frog's heart. The ventricle strip was stimulated electrically at 3-second intervals to contract, the contraction being reduced by acetylcholine. By ingenious calculations he determined the quantity of the agonist action upon each cell. The amount of drug that is fixed by the cell is very small: only 20,000 molecules fixed per cell are sufficient to produce a demonstrable action; this number of molecules could occupy only a very small fraction of its surface. He concluded that a reversible monomolecular reaction occurs between the drug and some substance either in the cell or on its surface [see Clark, 1985].

In 1887 the young Spanish Professor Santiago Ramón y Cajal proposed the structure of the brain with not much more than a simple light microscope, pen and paper, specimens of brain tissue and the tissue-staining method of the Italian histologist Camillo Golgi. Cajal suggested that the neurons receive information at their dendrites and cell bodies and relay nerve impulses through their axons. He stated that the axons ended and formed contacts (later termed synapses by Charles Sherrington) to dendrites and other nerve cell bodies. The unseen synaptic gaps were the basis of neurotransmission. In contrast to the 'reticular theory' of Golgi, Cajal tactfully defended his 'neuron doctrine' at their first meeting, when both received the Nobel Prize in Physiology or Medicine in 1906 [for a review, see Rapport, 2005].

One neuron sends information to another at the point of contact known as synapse. The main debate related to synaptic transmission was whether it was chemical by the release of neurotransmitters, or electrical like the nerve impulse itself. Physiologists believed that fast synaptic transmission had to be electrical. They used modern methodology and tools such as the cathoderay oscilloscope, whereas pharmacologists were still using bioassays and old-fashioned recording methods. But new technology finally convinced the physiologists that synapses do communicate by the release of neurotransmitters. John Eccles, one of the strongest proponents of electrical transmission in synapses, finally provided some of the key evidence that he, and the other physiologists, had been wrong. Eccles was a friend and great admirer of Dale [for a review, see Valenstein 2005].

A picturesque situation concerning the receptor concept has been described by de Jongh [1964]:

'To most modern pharmacologists the receptor is like a beautiful but remote lady. He has written her many a letter and quite often she has answered the letters. From these answers the pharmacologist has built himself an image of this fair lady. He cannot, however, truly claim ever to have seen her, although one day he may do so... and the beautiful lady soon appeared.'

In the 1970s, pharmacology entered a new phase. Receptors emerged as biochemical realities. The availability of tightly bound radioactive ligands made it possible to extract and purify the radioactively labelled receptor material [see Burgen, 1970]. This approach was first used successfully on the nicotinic acetylcholine receptor. Meanwhile, the molecular aspects of receptor-mediated mechanisms have advanced so far that they are dealt with in separate chapters in a modern textbook of pharmacology.

Transmitter substances and other types of mediator are the 'keys' by which the receptor is 'turned' into an active shape, to form the sensing element in the system of chemical communication that coordinates the function of cells.

Agonists and antagonists of transmitter substances, including their synthetic analogues, provided the pharmacological basis for their therapeutic application.

The chemical transmission of nervous impulses was the key to many discoveries in pharmacology and other fields of life sciences and medicine. This is specifically true for the discovery of receptors, and agonists and antagonists acting on them. The historical experiments resulted in facts, now to be found in textbooks. The neurochemical transmission, initially demonstrated at the peripheral terminals of efferent autonomic nerve fibres, has been found in all types of efferent and afferent peripheral nerve fibres, in the enteric nervous system and in the countless nerve connections of the central nervous system. The chemical neurotransmission can be regarded as the overall used chemical language, in which nerves talk to each other or to organs.

The initial findings centred around the experimental work of Henry H. Dale, Otto Loewi and W. Feldberg. Their historical findings opened the door to further investigations of neurotransmitters and extended the list of neurotransmitters to include further monoamines, amino acids, peptides, purines and nitric oxide.

The aim of this publication is to provide information about the scientists involved in the early discoveries, their experiments and their substances. Their documented memoirs, obituaries and photos, only supplemented by comments, should provide the reader with basic information about their personality, their scientific approach and their life at times which were anything but quiet and peaceful. The texts written by the scientists should also demonstrate to the reader how unexpected findings were sometimes the basis of a discovery.

Not only scientists have their history, but also the substances used by them, such as transmitters, their antagonists and other tools to influence their action. Modern textbooks of pharmacology primarily provide facts about drugs which are important for therapeutic use. The stories about these substances, from their discovery on, might also interest readers. They are, therefore, included in the second part of this book, containing some anecdotes and unusual events on the way to their therapeutic application.

References

Bernard C: La science expérimentale. Paris, Baillière, 1878.

Bowman WC, Rand MJ: Textbook of Pharmacology, ed 2. Oxford, Blackwell Scientific, 1980.

Burgen AS: The drug-receptor complex. Annu Rev Pharmacol 1979;10:7–18.

Clark DH: Alfred Joseph Clark 1885–1941: a Memoir. Glastonbury, C. & J. Clark, 1985.

de Jongh DK: Introductory remarks; in Ariëns EJ (ed): Molecular Pharmacology. New York, Academic Press, 1964.

Langley JN: On the reaction of cells and of nerve endings to certain poisons, chiefly as regards the reaction of striated muscle to nicotine and to curare. J Physiol 1905;33:374–413.

Rang HP, Dale MM, Ritter JM, Moore PK: Pharmacology, ed 5. Edinburgh, Churchill Livingstone, 2003.

Rapport R: Nerve Endings: the Discovery of the Synapse. New York, Norton, 2005.

Valenstein ES: The War of the Soups and the Sparks: the Discovery of Neurotransmitters and the Dispute over How Nerves Communicate. New York, Columbia University Press, 2005.

OTTO LOEWI

Otto Loewi on his 80th birthday.
Photo by Trude Fleischmann.

OTTO LOEWI – INTRODUCTORY REMARKS

I was introduced to Otto Loewi in Vienna in 1958, at the end of the Opening Ceremony of the 4th International Congress of Biochemistry, of which Loewi was Honorary President. He invited me to see him the next morning at the Hotel Regina (F.L). I did not expect him to keep me sitting at his side the whole morning, during which he was frequently interrupted by journalists and others wishing to welcome him. He wanted to know every detail about my research work. He told me many details about his experiments, and when I expressed admiration for his memory, he replied, 'The memory is a function of the interest' and added that he had not even retained copies of his own papers. I politely answered that I would try to collect all his publications. I did not expect to fulfil this promise until after my move to Tübingen. Eventually with the collaboration of the student Wolfgang Giere the book *Otto Loewi: Ein Lebensbild in Dokumenten* was published.

Loewi's life is documented by his autobiographic sketch published in 1960 (see pp. 11–29).

Henry H. Dale's obituary of Otto Loewi that had been published in the *Biographical Memoirs of the Fellows of the Royal Society* in 1962 (see pp. 30–53) one year after Loewi's death contains a complete bibliography. Giere and I had not been aware of this publication, since it had been distributed only to members of the Royal Society.

One year before he died Franz Brücke, Head of the Department of Pharmacology of the University of Vienna, contributed his recollections of Otto Loewi to a series covering Austrian Nobel Prize winners (see pp. 54–67). He described the years he spent with him in Graz, including stories about life in Vienna in those times and concluded with a poem by Otto Mayerhofer expressing an idea with which Loewi identified.

Stephan W. Kuffler's recollections, not reproduced in this book, reflect the warm feelings of a close friend of Otto and Guida Loewi, and focus particularly on the summer days spent in his beloved Marine Biological Station at Woods Hole, where Loewi felt truly at home and in his element. He was often seen sitting in the library, talking to young researchers [Lembeck and Giere, 1968, picture on p. 192].

Throughout his life Loewi took up every challenge with enthusiasm. Three episodes illustrate this aspect of his personality.

Archives in Strasbourg

When Professor Oswald Schmiedeberg provided Otto Loewi, then a freshman at the Department in Strasbourg, with details about an experiment he should perform, he concluded by saying that Loewi should look for further details in an article in volume 6 of the 'Archiv'. Schmiedeberg meant the *Archiv für experimentelle Pathologie und Pharmakologie*, which he had started jointly with the pathologist Klebs (Prague) and the clinician Naunyn (Königsberg) in 1873. As Loewi did not find the 'Archiv' in the laboratory he went to the City Library. The librarian was rather surprised and said that he could examine the requested volume only within the library. He returned with a large leather-bound volume, the archive of the City of Strasbourg from some time early in the 14th century. Loewi concluded that there had to be another 'Archiv' somewhere.

In the early years in pharmacology Otto Loewi became enthusiastic about some new findings in biochemistry. One of the consequences of this enthusiasm is described below.

'No, and Now You Leave Me Alone' Said the Professor to the 27-Year-Old Doctor

This event happened at one o'clock at night before beginning a series of experiments which resulted in the publication by O. Loewi: Über die Eiweisssynthese im Thierkörper [from Lembeck and Giere, 1968, pp. 84–85].

'As outstanding contributions I consider in the first place those which opened up new important fields of established methods which made possible such progress. Among the many papers of mine dealing with metabolism is one which quite generally has been considered as coming up to the definition of "outstanding" as just given. That paper was on "Protein Synthesis in the Animal Body" published in 1902.

Up to that time no one had succeeded in maintaining animals by feeding them in place of native proteins with the products of their degradation, even if the degradation had not gone very far. It looked therefore as if animals were not able to synthesise their own proteins from the degradation products. I knew of course about this state of knowledge. I was however not especially interested in that question. This state of mind, all of a sudden, was remarkably changed. One evening it happened that I read a paper just published by Friedrich Kutscher who was the biochemist at Marburg, my University. In that paper Kutscher showed that by long lasting incubation of an organ – he used the pancreas under antiseptic conditions at 35° – the proteins are decompensed to such a degree that there is no longer present any reaction characteristic of proteins. At once I got the idea and the firm conviction that I would succeed, by feeding, as the only nitrogen carriers, such degradation products of the whole organ, in reaching a protein synthesis in animals. If anybody would have asked me why I was so firmly convinced of the eventual success of such an experiment after all the negative experiments of the past, I could not have given any answer. I was so excited that, when I went to bed that night, I was unable to find any sleep. At one o'clock in the morning I got up again to make immediately sure whether Kutscher did not himself wish to perform these experiments. He was, of course, already asleep. I remember very well how he berated me for my brainstorm. Without opening the door, he yelled at me: "No, and now leave me alone." My project seemed so important to me that early next morning I sent my boss to him to verify that he really did not intend to work it out. His answer was: "I don't care whatever this fool does, who wakes up decent people after midnight."

I started the experiments at once. For a very long time I encountered the greatest difficulties, mostly because the dogs did not relish the unusual food, which in fact was not very appetizing. I am pretty sure that many others would have been discouraged and would have given up. I, however, had such strong faith in final success that I persisted. This was rewarded, as eventually I overcame all the difficulties and was able to prove the correctness of my hunch that animals are able to rebuild their proteins from amino-acids, their final degradation products. In fact, total degradation within the intestine has since been shown to be the normal fate of fed protein. From that time to the present day this discovery has proved to be of great importance for the science and practice of nutrition. At the time I made the discovery, I did not at all foresee such dividends, which happens often. It reminds me of a well-known saying of Benjamin Franklin. He was witnessing the first demonstration of a purely scientific discovery and people around him asked, "but what is the use of it?" Franklin answered: "What is the use of a new-born child?"

My discovery was soon verified by other scientists; it caused considerable commotion and helped me to establish me as a researcher.'

Comment on Otto Loewi's Work

Pharmacology did not initially attract his interest. And yet, he asked the famous Father of Pharmacology, Professor Oswald Schmiedeberg in Strasbourg, for a subject for his thesis. As a rule, people working in Schmiedeberg's laboratory were apprentices or competent scientists from all over the world; rarely students who had no preliminary knowledge or training.

When Loewi was appointed to the Chair of Pharmacology in Graz in 1909, at the age of 34 years, he had already a high reputation among pharmacologists, which is illustrated by a few examples. His Inaugural Lecture on January 25,1909, had the title 'Pharmacology and Clinic', which was presumably his intention at that time, but was followed by experimental work mainly in the field of physiology.

In his publications about diuresis, he found that the caffeine-evoked diuresis is explained by a vasodilatatory action on renal vessels (1905). Jointly with A. Fröhlich (1908, 1910) he discovered that following an injection of cocaine at a dose without an effect on its own, the subsequent effects of adrenaline on blood pressure, urinary bladder relaxation, salivary secretion and pupillary dilatation are potentiated; they found this about 50 years before the description of the mode of action of cocaine by Muscholl [1961].

In his initial years in Graz, before the series of 14 publications about neurotransmission between 1921 and 1930 started, he investigated together with Konschegg (1913–1918) the relation between the cardiac effects of digitalis and the calcium concentration on the isolated frog heart. Loewi wrote: 'With the fact that strophanthin causes a sensibilization for calcium, the question concerning its action is certainly not solved but narrowed; it means now: what is the mode of action of strophanthin by which the heart becomes more sensitive to calcium?' This finding has been confirmed by numerous experimental and clinical studies [reviewed by Repke, 1964].

As a professor in Graz, Loewi delivered carefully prepared lectures every day for 30 weeks of the year. At that time there was a rather limited number of medical students, who all had to be examined by him. This left him sufficient time for research under the rather limited circumstances in the poorly equipped laboratory. From his home he had to walk for 5 minutes to the department. Occasionally he took a coworker on a walk up to the Castle Hill. But since cars were not common at the time further excursions to the surroundings of Graz were rare.

The 'Napoleon Ash Tree' in the Garden

In the following I would like to mention an example of Loewi's historical awareness and his sense of humour. Close to the corner of the property on which Loewi built his 'Villa' there was a large ash tree under which Napoleon I stood in 1782 to observe his troops marching by. Loewi bought the small piece of land to preserve the historical tree. In the years 1810–1813 Louis Napoleon owned the neighbouring mansion which still exists; he was displaced to Graz by his brother Napoleon I. He enjoyed staying there and wrote:

> 'La ville des grâces au bord de l'amour',
> which might be translated as:
> 'A city of beauties ready for love',
> and which in German sounds like:
> 'The city of Graz on the bank of the
> river Mur'.

An event like this contributed to Loewi feeling happy in his environment.
All this came to an abrupt end in 1938.

Another Spontaneous Reaction

Another spontaneous reaction, typical of Loewi, occurred on April 30, 1954. He had received the following cable: 'I have the honour to inform you of your election today as Foreign Member of the Royal Society. Hinshelwood, Foreign Secretary Royal Society.' On the same day he sent a note to his son Guido:

'Now a word became true: wishes in the youth – one receives with richness in the age. Such an honour does not take the second place after the Nobel Prize Award, with consideration that the number of Foreign Members of the Royal Society is limited to 50.'

References

Lembeck F, Giere W: Otto Loewi: Ein Lebensbild in Dokumenten. Heidelberg, Springer, 1968.

Muscholl E: Effect of cocaine and related drugs on the uptake of noradrenaline by heart and spleen. Br J Pharmacol 1961;16:352–369.

Repke K: Über den biochemischen Wirkungsmodus von Digitalis. Klin Wochenschr 1964;42:157–165.

Dem trefflichen Kollegen
Lembeck
mit guten Wünschen
Onkel Otto

AUTOBIOGRAPHIC SKETCH

OTTO LOEWI

Reprinted with permission from The Johns Hopkins University Press: Loewi O., An Autobiographic Sketch. Perspect Biol Med 1960;4(1):3–25

My Childhood and Education

The background of a man's achievements is usually the most interesting part of his life story. An advantage of autobiography may be that its author reports more competently than others on his inner experience and its effects; yet, even so, fiction may come into play because a retrospective report may not always truly reflect the past as it happened.

Because of their fundamental and lasting influence on my intellectual and spiritual development I have to go back to the years I spent in the humanistic gymnasium in Frankfurt am Main, my native town. In my time, heavy emphasis was put on the study of Latin (nine years) and Greek (six years). We were made familiar with more than the grammatical subtleties of these languages. Through reading and commenting on original texts, we also became acquainted with the civilisation of classical antiquity, which covered almost the whole sphere of human aspirations. This kind of study had unique and lasting value, not so much by increasing factual knowledge, but by widening the horizon and encouraging the habit of independent thinking. At least equally important, it always seemed to me, is that during the formative years a young person spends a good part of study time on classical languages because they are not of immediate practical value and because the past has shown that this type of education usually favourably shaped the student's personality and influenced his whole attitude toward life. I was graduated from the gymnasium in 1891.

One of my dearest memories from those school years is of the place where until 1890 we regularly spent our summer vacations. It was my father's estate, which consisted of an old manor, a large, enchanting garden, and some vineyards. It was situated on a slope of the Haardt Mountains in the Palatinate, a wine district whose vintages have deservedly been as famous as those from the vineyards on the Rhine. Each time I went there with my family my feelings for our country place were deepened. My memories of each detail are as vivid as if they were from yesterday and not seventy years ago. The Haardt was where I first picked a lily of the valley and plucked my first apricots from a tree.

During the obligatory nine years that I attended the Gymnasium, I received poor marks in physics and mathematics. They were compensated by fairly good marks in the humanities. It was always understood that after graduation from the gymnasium I should study at a university. I was not one of those fortunate few who early display a definite talent which indicates the choice of study.

As a teenager I was already familiar with masterworks of the early Flemish painters through trips to Belgium. They deeply impressed me and caused me to become acquainted with works of art, particularly paintings, whenever possible and to read relevant literature. In fact, my leaning to art became so strong that I wished to study the history of art. The family, however, for practical reasons, was against this project and wanted me to study medicine. I gave in, and in the fall of 1891 went to Strassburg University. It ranked among the most excellent in Germany because after the Franco-German war Alsace-Lorraine had become German, and Strassburg, its university, received all kinds of privileges to attract the very best scientists as Professors.

After my matriculation as a medical student, I began to attend regularly all the required preclinical lectures and courses. Right then and there I found out that a great scientist is not inevitably a good teacher. Consequently, I soon limited my attendance almost exclusively to the anatomy courses of Gustav Schwalbe, who was both an excellent scientist and a stimulating teacher. A good part of the time gained by playing hooky from medical courses I spent attending lectures at the philosophical faculty – lectures covering the exact sciences as well as the humanities. I was particularly attracted by Georg Dehio, at that time the outstanding scholar of the history of German architecture. Another man who deeply impressed me was Wilhelm Windelband, a highly regarded philosopher; Kuno Fischer in Heidelberg, the most popular lecturer in philosophy, had Windelband in mind when in one of his lectures he remarked, 'There are only two philosophers in Germany. The other one teaches in Strassburg.' Only in the summer of 1893, during the fourth term of my medical study, did I seriously prepare for the first medical examination, called Physicum, to which one had to submit toward the end of that term. I was quite satisfied that I narrowly escaped failing.

The following year I spent in Munich. In the 1890's the revolution in literature which had been manifested first in the social and naturalistic plays of Scandinavian and German authors reached its acme. Berlin and Munich competed for the best performances. Munich excelled also in its performances of the operas of Wagner, who was at the zenith of his glory, and it had become the center of an exciting new movement in fine arts. These were the main reasons why I went to Munich. In the beginning I attended medical courses fairly regularly. But again my eagerness soon diminished, and I spent more and more time visiting the museums and exhibitions of what then was regarded as 'modern art' and on other nonmedical activities. After a year, in the fall of 1894, I returned to Strassburg. Only then, in the beginning of my fourth academic year, my indifference to medicine suddenly gave way to almost enthusiastic interest.

After signing up for the lectures, a student had to collect the signatures of the individual lecturers for his attendance record. One signature I needed was Professor Bernhard Naunyn's, the head of the clinic of internal medicine, an outstanding clinician and experimental pathologist as well as a stimulating teacher. His lecture was already in full swing when I arrived in the clinic. Even the corridor leading to the open door of the auditorium was filled to capacity. It was so quiet that, standing at the end of the corridor, I could catch every word of the speaker, whom, of course, I could not see. The sound and vigor of Naunyn's voice as well as his presentation of the subject so deeply impressed me that the next day I came half an hour early to be sure of a seat. From that day until the end of my studies I tried not to miss a single Naunyn lecture. I also volunteered for student practice because I wished to attend Naunyn's daily rounds in the wards and in the laboratory. In time I came into personal contact with this fascinating man and his excellent staff, among whom were Oscar Minkowski and Adolph Magnus-Levy.

Naunyn provoked my interest in several branches of medicine, but pharmacology was not one of them. And yet, strangely enough, I asked the famous 'Father of Pharmacology,' Professor Oswald Schmiedeberg, for a subject for my thesis. As a rule, people working in Schmiedeberg's laboratory were apprentices or competent scientists in the field from all over the world, rarely students who, like me, had no preliminary knowledge or training. What motive impelled me to this step? In the first part of Faust Goethe says, 'Du glaubst zu schieben und du wirst geschoben', or, as translated by Bayard Taylor, "Thou'rt shoved, thyself imagining to shove."

In the laboratory I again met Oscar Minkowski and also Arthur Cushny, Karl Spiro, and Walter Straub. Straub was still a student, but he had already been fairly well trained in experimental work in the laboratory of Carl von Voit, who was the teacher of Graham Lusk also. We became frieds for life. It was Spiro who introduced me to the classic studies 'On the Life of the Rhine Salmon in Fresh Water,' of Friedrich Miescher, Swiss pioneer in biology. These studies were collected and republished in 1897 as "The Histochemical and Physiological Studies of Friedrich Miescher." Time and again I have turned to this volume, which became to me a kind of scientific bible. While I owe my interest in medicine to Naunyn, Miescher's studies were chiefly responsible for my later choice of biology – strictly speaking, physiology – as a calling.

During the tenth term, I had to take my oral examinations (Staatsexamen). Everything went well until the last section of the last examination. The subject was surgery, and the last section included fractures and dislocations. I drew a question about interdental splints. I enumerated all of those referred to in the textbook, but the examiner inquired about yet another one. It was an obscure splint engineered by a dentist who had once worked with the Professor. This splint had failed to make the textbooks, and it was only natural that I never had heard of it. For this failure I was not passed for six weeks. In my opinion, most examiners who are feared by candidates because of their strictness, pettiness, and insistence on unimportant details are narrow minded and frustrated because of a hidden, often justified, inferiority complex. They are not so much critical of but find fault with generally acknowledged achievements. Naunyn once wrote, 'Criticism is indispensable, skepticism is the arrogant attitude of an ignoramus.'

After my graduation from Strassburg in 1896, my parents fulfilled an old desire of mine by offering me a trip to Italy. I have returned to that country time and again. I would not like to have missed the enrichment of my life that I owe to Italy.

I Choose the Laboratory Rather than the Clinic

I was fully aware that I lacked sufficient knowledge and training in chemistry to meet the requirements of experimental work in any laboratory. After my return from Italy, I therefore took an excellent course, in inorganic analytical chemistry in Frankfurt and then spent a few months in the biochemical institute of Franz Hofmeister in Strassburg. He was an ingenious biologist as well as an unforgettable personality. Meanwhile, Professor von Noorden, the internist of the city hospital in Frankfurt, offered me an assistant's position. Von Noorden was well known for his valuable studies on pathologic metabolism as well as on nutrition and dietetics. I had already decided to seek an academic career but did not know whether I should choose internal medicine or a basic medical science. Preparation for either branch of medicine by service in the ward combined with laboratory work appeared to me advantageous. I therefore accepted Noorden's offer and started in my post in 1897. The first ward I had to take care of was reserved for cases of far-advanced tuberculosis; in the second there were many patients suffering from a pneumonia which then was epidemic and proved to be particularly dangerous for vigorous young men. There was no therapy. The mortality was enormous. Because of these experiences, I decided not to become a clinician but to do research in basic medical science.

At that time, and ever since, I have looked at pharmacology as the science whose main goal is revealing physiological functions by the reactions of living matter to chemical agents. The very concept of using drugs in the study of functions oriented me to pharmacology.

On my request for advice about with whom I should work, Hofmeister strongly recommended Professor Hans H. Meyer, the pharmacologist at the University in Marburg an der Lahn. There was, fortunately, an opening in the department, and after an encouraging interview I was accepted as assistant. Dr. Meyer was a great scientist and person. His scientific reputation reached its culmination after the publication of his pioneering studies leading to his 'Theory of Narcosis' in 1899, and of his 'Researches on Tetanus' with Fred Ransom in 1903. In 1904 he was invited to deliver the first Harvey Lecture in New York. I stayed with Dr. Meyer for ten happy years, from 1898 until 1905 in Marburg, then in Vienna.

The conditions in Germany at that time were most favorable for the development of scientists. Inspite of the frequent rattling of sabers, under the reign of Wilhelm II almost everybody enjoyed both security and freedom. The universities were government owned, but the government hardly ever interfered with their far-reaching autonomy. As a rule, it contented itself with accepting the universities' proposals and paying the endowments and debts of the departments. If the debts were excessive, the department got a warning – and a little later the payment. Because of these enlightened policies, there was no need to rush work and publication. There was much leisure for keeping abreast of the literature – which, to be sure, was not so immense then as it is nowadays – for extensive discussions, and for concentrated thinking. I was not at all sure whether I had enough scientific imagination for a man who thought of spending his life in basic science. Therefore, at the interview with Dr. Meyer, I proposed and he agreed that my two first years with him should be on probation. Apparently, for this reason, Dr. Meyer encouraged me to make my own decisions about the subject matter of my investigations right from the first day. After I had made my decision, he discussed with me all of the details and then followed the progress of the work with great interest. His analytical talents and his imaginative mind were of inestimable value to me. So was his broad-mindedness and deeply rooted culture, which he expressed in digressions often lasting for hours into many aspects of human aspirations. In many respects I owe to Dr. Meyer a very large part of my education. During the first years in Marburg, I chose my topics mainly from the field of metabolism, toward which I had been drawn by personal contact with such pioneers in the field as Naunyn, Minkowski, and Magnus-Levy and by intense study of the fundamental papers of Miescher. I started by investigating the action of phlorhizin, a glucoside that provokes glycosuria. This was my first approach to the problems of carbohydrate metabolism, to which I returned time and again. Another investigation concerned nuclein metabolism in man. After the publication of the results obtained in this study, Dr. Meyer recommended me, in 1900, to the faculty as Privatdocent, a title signifying the first step in an academic career. I delivered the obligatory probation lecture on 'The Migration of Material in the Organism' because this subject gave me an opportunity to emphasize Miescher's pioneering work. In 1902 I published my paper on 'Protein Synthesis in the Animal Body.'

Up to 1901 no one had succeeded in maintaining animals in nitrogen equilibrium by feeding them with the degradation products of casein or fibrin in place of native protein, even if the degradation had not gone very far. It was therefore assumed that animals were not able to synthesize their proteins from degradation products. I knew about this state of affairs but was not especially interested in the question. My attitude changed remarkably overnight. One evening in 1899 I read a paper just published by the excellent biochemist Friedrich Kutscher, an associate of Albrecht Kossel in Marburg, 'On the End Products of Trypsin Digestion.' He showed that by prolonged digestion in an incubator, a pancreas can be decomposed so that any reaction characteristic of protein is no longer present. Right then I got the idea and the firm conviction that one would succeed in reaching protein synthesis in animals by feeding them the degradation products of a whole organ rather than of a single protein.

I started the experiments at once. For a long time I encountered great difficulties. Particularly because most of the dogs did not relish the unusual fare. I persisted, however, because I had not the slightest doubt that I would succeed in the end. My perseverance was rewarded. In 1902 I was able to prove the correctness of my hunch that animals are able to rebuild their proteins from their final degradation products, the amino acids. This discovery has played an essential part in the science and practice of nutrition. It also was the first publication that made my name known as a researcher. I did not at all foresee such dividends. In 1902 I also published the first part of a series of papers called 'Experimental Contributions to the Physiology and Pharmacology of Kidney Function.'

To England To Learn

The relatively simple techniques I applied in the experiments I had published, and those I prepared for Dr. Meyer's lectures on pharmacology were almost the only ones at my command. Frequently, because of my poor training in the methods required, I dropped ideas that occurred to me. This meant, of course, a serious limitation of my scientific activity in the future, and I decided to remedy this shortcoming.

During the nineteenth century the humanities as well as the basic sciences were blooming, especially in Germany. World-renowned German medical scientists like Julius Cohnheim, the experimental pathologist in Leipzig, and physiologists like Rudolph Heidenhain in Breslau and Wilhelm Kuehne in Heidelberg attracted students from abroad. But it was the laboratory of Karl Ludwig in Leipzig that became a kind of Mecca for physiologists and other medical scientists from all over the world. He devised innumerable methods and apparatus for approaching problems until then inaccessible. After Ludwig's death in 1895, Germany little by little lost its leadership to England, where men such as Barcroft, Bayliss, Gaskell, Haldane, Langley, Sherrington, and Starling had become world-renowned physiologists. Being familiar with most of their papers, I decided to spend some time in England to learn methods and to meet at least some of its famous physiologists. I wished particularly to work with Ernest Starling because his papers covered so many sides of experimental physiology. I wrote to him in 1902, and he accepted me as a student in his laboratory. On the trip to London I did not neglect to stop for a week in Holland to enjoy the original paintings of the Dutch masters which I had known until then only from prints.

When I first saw Starling's laboratory, I was surprised by the striking contrast between its extremely limited space and primitive equipment and the very high level of the work that had come out of it. Like everybody else, I was charmed at first sight by Starling's appearance, his expressive features, his shining eyes. I soon recognized that he was the most dynamic and contagiously enthusiastic man I had ever met, full of ideas and optimism, as well as very critical of other people's and his own achievements. He possessed a serenity and simplicity, humility and a kind of naiveté that is characteristic of so many men of genius. He made the atmosphere of his laboratory warm, informal, and stimulating.

W.M. Bayliss, Starling's brother-in-law, was also working there – an inconspicuous man, always in high spirits and extremely modest. His interest, knowledge, and thought covered almost every branch of science important for progress in biology. Aside from his broad view, he possessed all the other qualifications needed for writing that unique book 'Principles of General Physiology (1917)', which for many years had an enormous influence on everybody interested in the question 'what is life?' Many facts offered in this book are now obviously out of date, but it still makes stimulating and exciting reading, and for fundamental principles and historical perspective, nothing has yet taken its place. Starling and Bayliss were very generous in offering me so much of their wealth of ideas, knowledge, and experimental experience.

It was in Starling's laboratory in 1902 that I first met Henry Dale. We at once found that our interests and aspirations had much in common, and we soon developed a close, lifelong friendship.

I greatly enjoyed the meetings of the Physiological Society, an organization that impressed me as being one big family. The atmosphere of the meetings was informal, and a fine kind of humor spiced the discussions. I had never experienced anything like this in continental meetings. At these meetings I had the good fortune to become acquainted with most of the British physiologists of that time.

A friend, Nathaniel Alcock, who worked with me in Marburg and later succeeded Brodie as Professor of Physiology at the University of Toronto, conveyed to me an invitation to spend the Christmas vacation in the rectory of his uncle,

Dr. Kingsmill, in a community close to Norwich. The rector immediately excused me from attending the daily services in his home. On a Sunday morning, however, he said that his altar boy had fallen sick and asked me to substitute for him at the service in the local church. He carefully prepared and dressed me, and then – Otto Loewi, M.D., Privatdocent of Pharmacology in Marburg, functioned that day in a white vestment as acting altar boy.

During the stay in England, I came to admire the English people in every respect. This visit also left me with a warm feeling of friendship for individual Englishmen, beginning with the contemporary biologists and extending to the succeeding generations.

Vienna

After my return to Marburg, I continued to study the function of the kidney and the mechanism of action of diuretics with Walter M. Fletcher, Velyan Henderson, and Nathaniel Alcock. In 1904, Dr. H.H. Meyer accepted an appointment as Professor of Pharmacology in Vienna. I substituted for him in Marburg and after a year followed him to Vienna.

The seven years spent in Marburg were very enjoyable. I valued a close association with Hermann Cohen, the famous neo-Kantian philosopher, although I did not understand his philosophy. I was also on good terms and in stimulating scientific contact with many colleagues of my age and with Professors Friedrich Mueller, Rudolph Krehl, and Ernst Romberg, the successive heads of the outpatient clinic in Marburg, which position proved to be for them a steppingstone to brilliant academic careers.

In Vienna I made up for what I naturally had to forego in Marburg, which had just one per cent of the population of Vienna. I frequently went to the Hofoper, then under Gustav Mahler, and to the Burgtheater, where Joseph Kainz stood out; he is considered to be the equal of the greatest actors of all time. I indulged in the pleasure of studying art treasures accumulated during centuries and of becoming familiar with the unique charm of that city and its gracious surroundings. The Easter vacation of 1905 I spent in Naples and worked in the famous zoological station. There I became well acquainted with Anton Dohrn, its founder, the highly cultured friend of Adolph Hildebrandt, the architect, Joseph Joachim, the violinist, and Hans von Marées, the painter who had adorned the station library with frescos of world fame. It was there that I met for the first time Ernst Theodor Bruecke, who later held the chair of physiology in Innsbruck. From our first meeting we developed a close friendship which lasted until his untimely death in Boston, where his old friend Alexander Forbes had offered him opportunity to continue his work at Harvard.

The Viennese medical school enjoyed a high reputation of long standing that attracted many students, from the United States especially. There existed an 'American medical association in Vienna,' and Amerikanerkurse (Courses for Americans) were a standing Institution. A growing number of young scientists and clinicians, at first mostly Viennese, applied for admission as co-workers in the pharmacological department. With those who were assigned to me by Dr. Meyer, I again studied mostly problems connected with carbohydrate metabolism. One of these studies proved that the preference of pancreatectomized dogs for fructose rather than glucose, as had been demonstrated by Minkowski, is not specific for this deficiency but is shared by dogs deprived of their glycogen by other means, e.g., by phosphorus poisoning. It was further shown that the heart, in contrast to the liver, cannot utilize fructose. It was finally discovered that epinephrine injections into rabbits completely depleted by starvation of their liver glycogen brought the glycogen back to almost normal value in spite of continued starvation. Other investigations in Vienna were devoted to studies on the vegetative nervous system. I had imported the interest in that field from England, where Gaskell and Langley had discovered the existence of two divisions of the vegetative nervous system, and where T.R. Elliott during my short stay in Cambridge was just about to conduct the final experiments on the action of epinephrine. My classical paper on this subject was published in 1905. I did experiments in that field in Vienna jointly with Alfred Froehlich, well known for the syndrome named after him. A skilful experimenter and genuine musician, he was liked by everybody, particularly because of his serenity and equanimity. This gentle philosopher retained these qualities and developed them even more in 1938, when he was forced to leave his beloved Vienna. He went to the United States and adapted immediately to conditions of life very different from those he had been accustomed to in Vienna.

The best-known result of our studies on the vegetative nervous system was the observation that small doses of cocaine potentiate the responses of sympathetically innervated organs to epinephrine and sympathetic nerve stimulation. This reaction has proved to be so specific that it has been successfully used as a test for epinephrine or norepinephrine.

As I had done before, in 1907 I spent my summer vacation in the upper Engadin, in Switzerland. To me it has always been the most beautiful place in the world. During my stay in Pontresina, I was introduced to Dr. Guido Goldschmiedt, then Professor of Chemistry in Prague, later in Vienna, who vacationed in Pontresina with his wife and daughter, Guida. In 1908 I married Guida. She was my faithful companion, devoting her life to my well-being for over fifty years.

In 1909, I accepted the chair of pharmacology in Graz. It was with deep regret that I left Vienna after four stimulating and fruitful years.

Graz

Like Marburg, Graz is an old and picturesque town built around a hill known as the Schlossberg, which, incidentally, happens also to be the name of its counterpart in Marburg. Graz could count only one-tenth the population of Vienna, yet it was the second largest town in Austria and had fine concert halls, a playhouse, and an opera house. The performances well deserved their high reputation. As everywhere in Austria and Germany, the theaters in Graz were considered important educational institutions and therefore were maintained by the city rather than as private enterprises. We soon got on friendly terms with some of the many art and music lovers. During the winters of the twenties, we usually had monthly chamber-music concerts played in our home by friends who were professional musicians or excellent amateurs.

I remember well when and how I started to enjoy symphonic music. During my last year in the gymnasium, I was invited several times to the famous symphony concerts of my home town, but usually I could not help falling asleep. During my first year in Strassburg, a fellow student who was an outstanding cellist and music connoisseur encouraged me to accompany him to a few concerts. He also stimulated my interest in the structure of musical creations and helped me to understand something of their essence. Little by little and without a conscious effort I came to love music. We may possess hidden inclinations and qualifications which need only to be encouraged to be released and developed.

In Graz I had to deliver five lectures weekly during the whole long winter term. It took me more than two full years to prepare them. When a student became restless during a lecture, I always felt that it was not his fault but my own. Before and during the first minutes of a lecture or a talk, I suffered from a kind of stage fright. A similar kind of inhibition has apparently been the cause of my spending so much time and effort preparing every one of my publications and talks. It took me about six weeks to prepare my first independent paper for publication. When I apologized to Dr. Meyer for my long absence from the laboratory, it gave me some comfort when he answered, 'Having worked on your subject for almost two years, you should respect your work enough to devote six weeks to its write up.' Recently, New York's Park Commissioner Robert Moses published a short essay, 'A Toast to O'Casey.' Here one reads, 'If as the result of enormous effort and diligence the words seem to come naturally and with effortless ease, is not this the height of art?' I really appreciated this thought, since it precisely described what I have always been striving for.

I remember quite well that this strong self-criticism began when, in my early twenties, I became acquainted with Guy de Maupassant's short stories. I was fascinated not so much by the content of the stories, which I often disliked, but by the perfect shading of expression, and still more by the justification of each word used. Even slight details that at first seemed irrelevant proved in the course of the story to be necessary or essential. The economy of wording has ever since been to me one of the most important measures of quality in writing.

In Graz, before 1914, my first associate from the North American continent was the late Velyan Henderson, at that time Professor of Pharmacology in Toronto, who had worked with me in Marburg also. Next came Franklin C. McLean, who was then a professor in Portland, Oregon. After World War I, before leaving for the United States, Carl F. Cori came, followed by Harold G. Wolff, Ralph G. Smith, now at George Washington University, and John L.R. Browne and Rhoda Grant from McGill University. We also enjoyed welcoming as visitors from the United States John T. Edsall, Wallace O. Fenn, Herbert S. Gasser, Ralph W. Gerard, Ross G. Harrison, and Jeffries Wyman.

To have regular though modest physical conditioning, I adopted the habit of climbing almost daily to the top of the Schlossberg, usually in company of one of my co-workers. I always liked discussions with younger people. Since most of them are not yet preoccupied with scientific problems of their own, they are susceptible to other people's ideas and willing to listen.

Many studies in Graz were again devoted to problems connected with carbohydrate metabolism. Adolph Jarisch, a highly gifted scientist, who for many years was my assistant and then became professor in Innsbruck, elucidated the mechanism of action of Claude Bernard's piqûre. Then, with isolated rabbit hearts perfused with Locke solution, we studied the conditions responsible for epinephrine hyperglycemia.

Our last contribution to the field of carbohydrate metabolism concerned a particular kind of regulation. We called it a proprioceptive metabolic reflex. It demonstrates a mechanism through which the central nervous system, which is informed of the general metabolism by humoral means, is informed of the state of the metabolism in individual organs.

A biography should not vie with a bibliography for completeness. I, therefore, mention only one out of many studies on the part played by cations in some physiological functions and in drug actions. From the results of these studies, the first of which was fully published in 1913 and the last in 1944, I concluded that the effects of digitalis glucosides on the frog's heart are due to those drugs' sensitization of the heart to calcium. This theory became the starting point for quite a number of studies in various laboratories. Its application enabled me to elucidate the particular reaction of the toad's heart to digitalis, until then unexplained.

I had the opportunity to give a short account of the first results obtained from these studies in Dundee when I was invited to attend the annual meeting of the British Association in 1912. Among the entertainments offered at the meeting was a trip to Glamis Castle, well known from Macbeth, owned since then by the Earl of Strathmore. When the visitors were gathered on the lawn surrounding the castle, one of the bagpipe players – to my greatest astonishment – shouted my name. He then led me to Lady Strathmore, the grandmother of Queen Elizabeth II. Lady Strathmore graciously guided me to many remarkable sites in the castle. After the tour I asked Starling how it happened that I alone was extended this privilege. He told me that Lady Strathmore had once mentioned to one of his friends how bored she used to be with such collective visits, because out of shyness the visitors never dared to talk to her. Starling immediately recommended me as one who would definitely not suffer from such shyness.

Following the meeting, Starling, to my great pleasure, joined me in a trip through the famous lochs in Scotland. Once before, in 1904, we had traveled together in the Austrian and Bavarian Alps. He then looked so keen and vigorous that the mountain guides asked me whether he was a guide I had brought along. Starling, who usually rounded Regent Park at a trot every day before dinner, climbed in one day what for others took two days. No guide was willing to go with him after the first excursion.

The idea of Chemical Transmission of Nervous Impulse

Now I have to turn to the best known of my scientific achievements, the establishment in 1921 of the chemical theory of the transmission of the nervous impulse. Until 1921 it was generally assumed that transmission was due to the direct spreading of the electrical wave accompanying the propagated nervous impulse from the nerve terminal to the effector organ. Since the character of that potential is everywhere the same, such an assumption would not explain the well-known fact that the stimulation of certain nerves increases the function of one organ and decreases the function of another. A different mode of transmission had, therefore, to be considered.

As far back as 1903, I discussed with Walter M. Fletcher from Cambridge, England, then an associate in Marburg, the fact that certain drugs mimic the augmentary as well as the inhibitory effects of the stimulation of sympathetic and/or parasympathetic nerves on their effector organs. During this discussion, the idea occurred to me that the terminals of those nerves might contain chemicals, that stimulation might liberate them from the nerve terminals, and that these chemicals might in turn transmit the nervous impulse to their respective effector organs. At that time I did not see a way to prove the correctness of this hunch, and it entirely slipped my conscious memory until it emerged again in 1920.

The night before Easter Sunday of that year I awoke, turned on the light, and jotted down a few notes on a tiny slip of thin paper. Then I fell asleep again. It occurred to me at six o'clock in the morning that during the night I had written down something most important, but I was unable to decipher the scrawl. The next night, at three o'clock, the idea returned. It was the design of an experiment to determine whether or not the hypothesis of chemical transmission that I had uttered seventeen years ago was correct. I got up immediately, went to the laboratory, and performed a simple experiment on a frog heart according to the nocturnal design. I have to describe briefly this experiment since its results became the foundation of the theory of chemical transmission of the nervous impulse.

The hearts of two frogs were isolated, the first with its nerves, the second without. Both hearts were attached to Straub canulas filled with a little Ringer solution. The vagus nerve of the first heart was stimulated for a few minutes. Then the Ringer solution that had been in the first heart during the stimulation of the vagus was transferred to the second heart. It slowed and its beats diminished just as if its vagus had been stimulated. Similarly, when the accelerator nerve was stimulated and the Ringer from this period transferred, the second heart speeded up and its beats increased. These results unequivocally proved that the nerves do not influence the heart directly but liberate from their terminals specific chemical substances which, in their turn, cause the well-known modifications of the function of the heart characteristic of the stimulation of its nerves.

The story of this discovery shows that an idea may sleep for decades in the unconscious mind and then suddenly return. Further, it indicates that we should sometimes trust a sudden intuition without too much skepticism. If carefully considered in the daytime, I would undoubtedly have rejected the kind of experiment I performed. It would have seemed likely that any transmitting agent released by a nervous impulse would be in an amount just sufficient to influence the effector organ. It would seem improbable that an excess that could be detected would escape into the fluid which filled the heart. It was good fortune that at the moment of the hunch I did not think but acted immediately.

For many years this nocturnal emergence of the design of the crucial experiment to check the validity of a hypothesis uttered seventeen years before was a complete mystery. My interest in that problem was revived about five years ago by a discussion with the late Ernest Kris, a leading psychoanalyst. A short time later I had to write my bibliography, and glanced over all the papers published from my laboratory. I came across two studies made about two years before the arrival of the nocturnal design in which, also in search of a substance given off from the heart, I had applied the technique used in 1920. This experience, in my opinion, was an essential preparation for the idea of the finished design. In fact, the nocturnal concept represented a sudden association of the hypothesis of 1903 with the method tested not long before in other experiments. Most so-called 'intuitive' discoveries are such associations suddenly made in the unconscious mind.

Many questions connected with and raised by the discovery of chemical transmission were studied in laboratories all over the world as well as in my laboratory. We found, for instance, that the effect of the Vagusstoff on the heart quickly fades because it is inactivated by an ester-splitting enzyme (cholinesterase).

The Vagusstoff was soon identified as acetylcholine. Also, it was proved that the activity of cholinesterase is prevented by the alkaloid physostigmine. This was the first identification of the point of attack of an alkaloid and, to my knowledge, the first elucidation of the mechanism underlying all the effects of an alkaloid. Not until 1936 could I identify the transmitter liberated by stimulation of the accelerator nerve as epinephrine.

In 1926 the Committee of the International Congress of Physiology to be held that year in Stockholm requested me to demonstrate there the experiment on the frog heart. I did not feel too happy about this invitation. Like most experimenters, I had experienced time and again that experiments before a large audience often failed although they never did in the rehearsals. Fortunately, I was able to demonstrate in Stockholm the experiment not less than eighteen times on the same heart.

In 1928, I had the honor to be invited by the Royal College of Physicians in London to attend the tercentenary of the publication of William Harvey's book De Motu Cordis. The privilege of offering the first banquet at that ceremony was given to the 'Worshipful Company of the Grocers.' This ancient guild had once given Harvey a grant that enabled him to study at continental universities. As one of the ceremonies during the banquet, a silver horn filled with wine was passed around. Each time a guest drank from this horn his neighbor would rise and guard his back. This custom was instituted after a king of old was stabbed in the back while drinking.

Something else happened on that occasion that I never will forget. One of the foreign guest speakers thought this banquet a fitting opportunity to lodge the claim that one of his countrymen, and not Harvey, had been the first to discover the circulation of the blood. I became so angry about what seemed to me to be an unheard of absence of good taste that I shook my head and turned around expecting to see my English hosts react with similar indignation. How astonished I was to see all my English friends leaning back comfortably with broad smiles on their faces. With typical English equanimity, they considered the apparently insulting attitude of the speaker just a huge joke.

In 1933, I delivered the Dunham lectures at Harvard. During the time I stayed in Boston, my friendly relationship with Walter B. Cannon, which had begun at the International Congress of Physiology in Boston in 1929, developed into a warm personal friendship. My admiration for Cannon's human qualities matched my admiration for his scientific thought and achievement.

In 1935, I gave the Ferrier lecture to the Royal Society of London, with Frederic Gowland Hipkins in the chair. Guida and I greatly enjoyed Sir Henry and Lady Dale's hospitality. It was in their home on a rainy day that I came across Bayard Taylor's translation of the first part of Goethe's Faust. This translation follows the original almost to the letter and seems to have lost hardly any of the beauties of the original.

The Nobel Prize

In 1936, jointly with my dear old friend Dale, I was awarded the Nobel Prize in physiology and medicine. Twice during that unforgettable ceremony in Stockholm I was deeply moved. Heralded by trumpeters, the Laureates of that year made their entrance upon the stage, and at that moment, led by the almost eighty-year-old King Gustav, the whole audience rose to pay homage to science. Then after the prize was handed to Dale and me, the orchestra played the Egmont overture. The orchestra was located on the top gallery. To me the music seemed to come from heaven.

On the way back to Graz I stopped in Vienna to meet Sigmund Freud for the first time. I cannot better or more exactly describe my impression when alone with him in his studio than to say that he filled the whole room with his personality. I had a similar feeling with Joseph Kainz, the actor, and again when I met Einstein. These three figures were all men of genius, unassuming and even humble. People of this kind – of course, not all of them – take their achievements for granted.

1938

On March 11, 1938, I conducted with one of my associates an experiment that was supposed to be the last of a series which had all shown that afferent nerves, in contrast to efferent nerves, contain not even traces of acetylcholine. In the late afternoon my co-worker told me that the Nazis had just taken over the country. I was so strongly preoccupied with thoughts about our most recent finding that I did not catch the full import of that news! This almost incredible indifference still lasted while, before going to bed, we heard over the radio the deeply moving valedictory address of Schuschnigg as President of Austria.

I awoke from a deep sleep at three o'clock in the morning when a dozen young stormtroopers, armed with guns, broke into my bedroom, took me downstairs, and pushed me without any explanation into a waiting prison van that took me and others to the city jail. Later that night I was joined there by our two youngest sons, Victor and Guido. Our two elder children, Hans and Anna, were not in Austria at that time and, of course, did not return.

During the obligatory morning walks in the prison yard, I found that hundreds of other male Jewish citizens of Graz shared my fate. To increase the hardships of the jail, playing any game, reading, or writing were strictly prohibited.

When I was awakened that night and saw the pistols directed at me, I expected, of course, that I would be murdered. From then on during days and sleepless nights I was obsessed by the idea that this might happen to me before I could publish my last experiments. After repeated requests, a few days later I was permitted to have a postal card and a lead pencil to write, in the presence of a guard, a communication of my last experiment to be sent to 'Die Naturwissenschaften'. Later in the day the guard came to tell me that he himself had mailed my card. I felt as relieved as if our whole future depended on that communication. After two months I was released from the jail, and in three weeks my boys were set free and could emigrate.

Forced by the Nazi authorities to leave Austria, I departed from Graz on September 28, 1938, for London. Before leaving, in the presence of Gestapo men, I had to order the Swedish bank in Stockholm to transfer the Nobel prize money, deposited with the bank in 1936, to a prescribed Nazi-controlled bank. Guida was forced to remain in Austria because the Nazis were dickering with the Italian government to dispossess her of some real estate in Italy which had been in her family for over one hundred years. Apart from this, the Nazis deprived us both of all our property and belongings down to the last penny. After a pauper's oath was presented, Guida was given a so-called 'subsistence allowance.' It was not until 1941 that she could leave Austria and join me in the United States.

A few weeks after my arrival in London I accepted an invitation of the Franqui Foundation to serve for eight months as a visiting professor at the Université Libre in Brussels. To my great pleasure, simultaneously Erwin Schroedinger and Albert Szent-Gyorgyi became Franqui professors, the former in Ghent, the latter in Liège. The late Leo Zunz, an especially kind-hearted man, head of the Department of Pharmacology at the Université Libre, gave me the opportunity to work, and Walter Dulière, a biochemist, immediately offered me his services as an associate. He proved to be a great help in my laboratory work as well as in the preparation of the lectures which I had to deliver in French. He had full command of the English and German languages and literature as well and a deep understanding of the respective cultures.

In contrast was another experience I had in Brussels, with a charming lady, an acknowledged poetess. I was struck that she did not know even by name Gottfried Keller and Conrad Ferdinand Meyer, Swiss writers whose poetry and prose, written in German, were considered classics in German-speaking countries. I was still more surprised that the lady did not know by name Timmermans, a Belgian author, born and living in Lier, a village quite close to Brussels. His excellent novels and short stories, written originally in Flemish, were translated into German though not into French. I became conscious for the first time of the barriers presented by language. This is most unfortunate, particularly when there is ever increasing striving after understanding of foreign nations. Are the great writers not so representative of the character and level of civilization of a nation as the great artists are, or even more so?

During my stay in Brussels, I enjoyed much pleasant hospitality. I remember with deep gratitude Mrs. Max Gottschalk, who, through enormous effort, succeeded in getting visas for my daughter and her family, thereby enabling them to leave Czechoslovakia in time.

A New Citizenship

In the summer of 1939, I spent some months at the country seat of friends in England. Then, because of the outbreak of the war, I had to abandon my plan to return to Belgium and went to Oxford on invitation of Professor Gunn, who held the chair of pharmacology at the Nuffield Institute. While there I received a call from New York University Medical School to join the faculty as a research professor of pharmacology and to work in the laboratory of George Wallace, unforgettable to everybody who was fortunate enough to meet this distinguished scientist, teacher, and friend.

When I appeared before the American Consul in London in order to get my non-quota visa, he said that everything was ready for me except for a little formality. He asked whether I could prove that I had been teaching during the last few years. I produced the decree by which I had been dismissed as professor from the University of Graz, but he shook his head and said that this would only prove that I had been dismissed, not that I had been in a teaching position. I suggested that he should ring up Sir Henry Dale, who would be able to give him any information about me. He rejected this proposal as he considered Sir Henry to be just a private person. He was quite at a loss what to do because he really wanted to help me. Finally, quite casually, I mentioned that, of course, he would find all my data in Who's Who. 'You think you are in?' 'I know I am.' He sent for the book and felt quite relieved when he said that everything was O.K. Then I had to pass a physical examination and was eventually handed a sealed envelope by the doctor. After this procedure, I received my visa. I thanked the consul for the trouble he had taken on my behalf. When I reached the door, I turned again and said, 'By the way, do you know who wrote this paragraph for Who's Who? I did. Good-bye, Mr. Consul.' Upon my arrival in New York harbor, a clerk prepared my papers for the immigration officer. While he was busy doing this, I glanced over the doctor's certificate – and almost fainted. I read. 'd Senility, not able to earn his living.' I saw myself sent to Ellis Island and shipped back to Mr. Hitler. The immigration officer fortunately disregarded the certificate and welcomed me to this country. I arrived here June 1, 1940. On April 1, 1946, I became an American citizen.

I am one of those lucky people from abroad who right from the outset could continue here the work they had to abandon in the country they had considered to be their homeland. I owe this privilege to the understanding of my situation, the broad-mindedness, and the generosity of New York University College of Medicine as represented by Deans Donal Sheehan and Currier McEwen as well as to the late George Wallace. The unique climate of the school and its high reputation are in great part the product of a strong feeling of responsibility and a disinterested striving of the individual faculty members to serve the school.

I have been an active member of the staff since 1940, when I was sixtyseven years old. Until 1955 I continued to work experimentally and to publish mostly on leftovers from earlier problems. In 1944 I received the Cameron prize from the University of Edinburgh and an honorary science degree from my new school; and in 1950 the same degree from Yale University. Also in the 1950's I received honorary degrees from the universities in Graz and Frankfurt.

In 1954 I became a foreign member of the Royal Society of London. Frequent invitations to deliver lectures gave me the welcome opportunity of becoming acquainted with many excellent biologists and their activities as well as with a good part of the country. An ardent desire to see the West was met by an appointment in 1942 as Walker Ames Visiting Professor at the University of Washington. On the way back to the East via California, we stopped at some places because of the uniqueness of their beauty or because I had to lecture. At the California Institute of Technology I had the honor to be introduced to the audience by the late Thomas Hunt Morgan. In 1948, I spent three weeks as visiting scholar of the Richmond Area University Center. I had to lecture at sixteen universities and colleges, and thus we became acquainted with Virginia. The originality of Thomas Jefferson's architectural and layout plans made the University of Virginia, to me, the most outstanding of all the public buildings I have seen so far in this country.

In July, 1940, a month after my arrival in this country, I went to Woods Hole, Massachusetts, for a summer vacation. And since then until three years ago, Guida and I went there every summer, as do so many who once experience the unique atmosphere radiating from the Marine Biological Laboratory, which attracts renowned scientists from Europe as well. The atmosphere originates in the main from the opportunity for informal mutual interchange with broad-minded scientists and carefully selected, refreshingly curious students. Ever since my first day at Woods Hole, I have admired the high level of those students' discussion with their teachers. I myself felt as a student there. In fact, I owe to Woods Hole my introduction into quite a few areas of biology and the intensification of may interest in any aspect of cell structure, cell function, and their interrelations. Also in Woods Hole, I had the opportunity to keep abreast of the most recent progress in diverse branches of biology by attending many lectures given by competent scientists and by frequenting the library, considered to be the world's greatest of its kind. I met there old friends and made many new ones. As a matter of course, from the start I have felt at home in Woods Hole.

> Caelum, non animum, mutant
> Qui trans mare currunt. – Horace, Epistle I
>
> Who rushes oversea will find
> The climate changed but not his mind.

Like the great majority of those who succeeded in leaving Hitler-dominated countries in time, during the first years we were not able to maintain the standard of life we were accustomed to. However, it took us almost no time to adjust to the different pattern of life. In fact, it appeared to us an almost negligible inconvenience, far outweighed by the luck we were blessed with every single member of our family was able to leave Hitler's part of Europe. Furthermore, my strong desire to continue my educational and scientific activities was fulfilled by my appointment to New York University and the opportunities given to me by the Marine Biological Laboratories at Woods Hole and the many other institutions where I gave lectures or addresses.

It was good luck indeed that I got an appointment in New York. At no other place in the world could I have met and become acquainted with so many outstanding biologists from all over the world and other renowned people. It is a truism to say how much more complete our picture of people we know only through their achievements becomes when we meet them, even if only once. New York, furthermore, excels by having a peculiarly stimulating climate – meant, of course, only figuratively – which appears to be the source of permanent, quick progress in every respect. During the relatively short time I have lived here, I have been privileged to experience a spiritual growth of this city which manifests itself in all spheres of cultural life. For instance, there has been a tremendous increase in musical activities and a remarkable deepening of understanding on the part of the people. And I have observed the overwhelming response of young and old to the admirable opportunities offered to them in the museums. These examples could, of course, be multiplied indefinitely.

Until the Hitler catastrophe, I felt perfectly happy during the almost thirty years I spent in Graz. In fact, everything I needed or desired for my scientific and educational activities, including enough excellent co-workers, as well as a fairly large, interested audience, was there. Because of the age limit, even under normal political conditions I would have been obliged to metamorphose from an active to a retired professor. As such, one frequently is invited or allowed to stay as a permanent guest in a corner of an institute. It is quite understandable that, as a rule, the retired professor becomes to his host just as welcome as the mother-in-law to a young married couple. Anyway, my curiosity and receptivity so far have not diminished. Therefore, I am happy and deeply grateful to the fate that transported me to this country, my new homeland, and to this unique city, where I continue to enjoy the stimulating, almost rejuvenating influence in new friendships and the wealth of new impressions and experiences.

OTTO LOEWI 1873–1961

H.H. DALE

Reprinted with permission from the Biographical Memoirs of Fellows of The Royal Society 1962;8:67–89

OTTO LOEWI, born on 3 June 1873, in Frankfurt am Main, was the son of Jakob Loewi, a wine-merchant of that city, by his wife, Anna Willstaedter. Throughout his life Otto retained very happy memories of his childhood and schooldays. For the nine years from 1881 to 1890 he attended in Frankfurt a Gymnasium of the old style, in which the studies were centred on the classical languages – Latin being studied for the whole nine and Greek for the six later years. He accepted for himself the not unfamiliar view that such an emphasis in education, during the formative years of boyhood, would have had a specially favourable influence on his personality, and on his general attitude to life and learning. In an 'autobiographic sketch', published in the year before he died he recorded that, during these nine years at the Gymnasium, he obtained 'fairly good marks' in these literary subjects, but 'poor marks' in physics and mathematics. Without attempting to assess the respective shares in the result, of inborn aptitudes and of a particular educational routine, it may at least be accepted as a fact, that Loewi exhibited in later life an unusually wide range of knowledge and interests, for one whose main career and activities were concerned with one faculty of the natural sciences. There were, in any case, cultural influences at work on him during these years of his boyhood, other than those attributable to any part of his formal schooling. In addition to his residence in Frankfurt, his father had a small country estate in the Palatinate, on a slope of the Haardt Mountains, with an old manor house, a large garden and some vineyards; and Loewi retained enthusiastic memories of the holidays spent annually, during his schooldays, in surroundings which were so attractive and stimulating to a young imagination. And he recalled other visits during the same period to Belgium, and a life-long enthusiasm which he therewith acquired for the achievements of the early Flemish Masters of painting.

It is perhaps not surprising then to learn from his recollections that, when the time came for matriculation at a University, and for choosing the course to be followed there, his own preference would have been for the history of art. His parents, on the other hand, presumably with a view to a safer prospect of income, urged him to qualify for a career in medicine; and he eventually accepted their advice.

On his own confession, however, though he duly matriculated at Strassburg as a student of medicine, he found little to attract him in the formal lectures dealing with its preclinical subjects. Schwalbe's teaching of anatomy was exceptional in holding his interest, and otherwise he seems to have spent a good part of his first year in attending lectures on the history of architecture and on general philosophy; with the result that he narrowly escaped failure in his first medical examination ('Physikum'), at the end of his fourth term. And then he spent a year at the University of Munich, where his medical studies appear again to have encountered a serious rivalry from the opportunities which that city offered for new artistic experiences – dramas from Scandinavia, Wagner's operas, and the treasures of its museums and galleries.

It is, of course, not improbable that Loewi's memory – like that of many of us – would have been apt to give an undue prominence to these distractions from the prior demands of his chosen course of study, when he recalled the happy freedom of his student years. He had certainly acquired a wide and vigorous interest in all aspects of the culture and the artistic heritage of Europe. On the other hand, he responded readily to personal and scientific distinction in his teachers; and, on his return to Strassburg in 1894, for his fourth academic year, a new enthusiasm for his medical studies was awakened by contacts with such eminent leaders as Naunyn in clinical medicine, Schmiedeberg in pharmacology, and Hofmeister in biochemistry. The strongest impression appears to have been made on him by Naunyn's lectures and clinics; on the other hand, it was from Schmiedeberg that he sought a subject for the thesis which he had to present before graduation, and with it the opportunity for a first experience of practical research in the laboratory. Loewi could not recall any special reason for this choice, and, rather characteristically, explained it by a quotation from Goethe's Faust – 'Du glaubst zu schieben und wirst geschoben.' The subject which Schmiedeberg suggested – a quantitative measurement of the activities of hydrocyanic acid, arsenic and phosphorus on the isolated heart of the frog – hardly seems likely, in retrospect, to have excited a lively interest in pharmacology; but it was presumably intended to introduce the student to the methods then in current use. On the other hand, Loewi gained much from his contacts with other men then work – in the department – Oscar Minkowski, Arthur Cushny, Karl Siro and Walter Straub – with all of whom he was able to share new interests and to make more lasting friendships. It was through Spiro that he came to know, and to acquire a life-long enthusiasm for Friedrich Miescher's comprehensive studies of the changes in the metabolism of the salmon in the Rhine, during the fresh-water phase of its life. A glance at the bibliography, at the end of this Memoir, will make clear, indeed, the frequency with which Loewi's investigations, during a large part of his career, were to deal, often recurrently, with one or another of certain essentially metabolic problems, which had attracted and retained his interest. From a study of his publications, indeed, one might get the idea that a stronger educational background in chemical theory and practice might well have led him to seek his main scientific opportunity in biochemistry, rather than in pharmacology. He was aware, indeed, of the gap in his equipment; and after he had graduated at Strassburg, and had enjoyed a long holiday in Italy which his parents then kindly arranged for him, he took a course of chemistry in Frankfurt and returned to Strassburg to work for some months under Hofmeister's biochemical stimulus.

1897–1902

The time had come, however, in 1897, for Loewi at least to make trial of some more definite line of activity in medicine; and it seems that the interest which had been awakened by Naunyn's clinical teaching was still strong enough to lead him first to accept an assistantship in the medical department of the City Hospital in Frankfurt, which had been offered to him by Professor von Noorden, who had specialised in his practice on nutritional disorders and dietetics. Loewi soon found the experience discouraging, however, arid decided to seek an opening to a career offering opportunity for experimental researches, in more fundamental aspects of medical science. It seemed that he might find this in pharmacology, which had so far been largely concerned with drugs as agents for the experimental analysis of physiological functions, rather than with the discovery of a scientific basis for their rational uses as remedies. The modern closer association of pharmacology with practical therapeutics, and with the experimental discovery of new and more specifically effective remedies for the causes or the symptoms of different diseases, was then still many decades ahead; and Loewi, in any case, was to find his main pharmacological interest in the use of drugs to explore the nature of physiological activities, by studying the ways in which they specifically strengthened, reproduced, or interfered with these. It was, indeed, the study, on such lines, of the actions of certain alkaloids and natural bases, which was eventually to provide the clue to the best known and the most widely influential of all Loewi's scientific enterprises and discoveries. Having decided, then, to try his hand further in pharmacology, Loewi, on Professor Hofmeister's advice, enquired in 1898 whether Dr Hans Horst Meyer, then the Professor of Pharmacology at Marburg a.d. Lahn, could offer him an opportunity; and he was fortunate in finding that Meyer had a vacancy for him. In the event he was to remain with Meyer's department, as Assistant, Privat-Dozent and Associate Professor, for twelve years in all – seven at Marburg and then five more at Vienna, when Meyer was called to the Chair of Pharmacology there. Loewi soon acquired, and always kept, a deep respect for and devotion to Meyer, as a distinguished scientist, a fine personality and a trusted counsellor.

As will be seen again from the bibliography, Loewi's researches during his earlier years at Marburg, until 1902, were chiefly concerned with biochemical problems of metabolism, though a pharmacological interest was given to some of them, by the use of phlorhizin to produce an artificial diabetes, and by a study of the effect of camphor on the condition thus produced. Already in 1898 he had described a urea-forming enzyme in the liver; and in 1900 came several publications on the metabolism of nuclein. In 1901 he was tackling the question whether the animal body could form sugar from fat, and reaching the conclusion that it could not; and then, in 1902, came what will surely be regarded as the most important discovery which he made, at the early stage of his independent activity in research. Nobody had so far succeeded in maintaining a nitrogenous equilibrium in the nutrition of an animal, when the amino-acid products of a complete hydrolysis were substituted for natural, unbroken proteins in the food; and the conclusion had till then been generally accepted, that an animal's diet must contain proteins which had been, at most, only partially hydrolyzed by digestion, if it was to be able to build from them those which it required for the maintenance of its own tissues. Kutscher, however, had then recently shown that it was possible to disintegrate the proteins of an organ completely, into a mixture of their constituent amino-acids in their natural proportions, by a sufficiently prolonged tryptic digestion. Loewi followed this clue; and, training his experimental dogs to accept such a mixture as the sole nitrogenous component of their diet, he had succeeded in maintaining them in nitrogenous equilibrium. I remember the late Sir Frederick Gowland Hopskins speaking to me with some enthusiasm of the importance of this discovery. It seems obvious, indeed, that it may have contributed to the background of his own decision, to test the effect on maintenance and growth of feeding rats on all the then known constituents of a normal diet, each in a state of chemical purification – a decision which, again, was to furnish the basis for his own leading part in the discovery of the 'accessory food factors', later to be known as 'vitamins'.

Visit to England

The occasion for Loewi's first meeting with Hopkins was provided by a few months' visit which he made to England, in 1902. He wished to widen his experience of physiological technique, and had decided that England could at that time offer him better facilities for such a study of methods than were to be found in Germany, since the death of Carl Ludwig. He applied to the late Ernest Starling, the Professor of Physiology at University College, London, for the opportunity which he had in view; and this was readily offered. Loewi was naturally attracted by the dynamic quality of Starling's character, his buoyant leadership and his contagious enthusiasm for his own subject of physiology; also by the gentler personality, the natural modesty and the wide scientific erudition and insight of Starling's brother-in-law and frequent collaborator, the later William M. Bayliss. Loewi arrived at a time of vivid interest and exciting progress in the work of the laboratory at University College; for it was earlier in the same year, 1902, that Bayliss and Starling had discovered the hormone mechanism – by the liberation of 'secretin' and its carriage in the circulation – by which the presence of acid in the duodenum excited the secretory activity of the pancreas. The actions and properties of secretin, and of the 'depressor substance' which accompanied it in artificial extracts from the duodenal mucosa, were also providing various opportunities for experimental enquiries by others working in the laboratory – myself among them, holding then a George Henry Lewes Studentship.

Loewi was immediately and easily on terms of good fellowship with all of us there; and my own brief encounter with him was to develop later into a close friendship and a sharing of interests, for nearly 60 years. Others whom I remember as enjoying there the contacts with him, were W.A. Osborne, who was soon to depart for a Chair of Physiology in Australia, and F.S. Locke – a gifted amateur of physiology, who, in one of his active periods, had enriched its technique by his method of maintaining the activity of the isolated mammalian heart, perfused with his appropriate modifaction of Ringer's saline solution. Loewi's visit was to be a short one, and he had no plan to begin any research of his own during his stay. He sought the opportunity for acquiring experience by watching English physiologists at work, and discussing with them their methods, plans and ideas. He was then, indeed, as always, a voluble but interesting talker. One of the objects of his visit, we understood, was to acquire some conversational facility in English; but we soon learned that he was impatient of correction in his use of it. 'No', he would say, 'I have not time to learn English correctly; I wish to speak it fast.' And I find that some of his achievements in that line have had a legendary survival, being still known even to some English physiologists of the present Generation, 60 years later. I remember meeting him on his return from a short visit to Oxford, where he had been welcomed by the physiological circle; and he had naturally felt greatly honoured when he was told that the veteran physiologist, who had then become the Regius Professor of Medicine, Sir John Burdon Sanderson, had expressed a special wish to meet him. It was characteristic of Loewi's buoyant reaction to a new view of a situation, that he seemed to find a further cause for pleasure, though hardly of the same kind, in discovering that Sir John's desire to meet him had been mainly due to a wish to hear Loewi's improvised English, the rumour of which had already reached Oxford.

Loewi, indeed, had easily found himself at home in the friendly informatility of the English physiological circle, which extended to the community, then a comparatively small one, of the Physiological Society; and he was ready then, and later in life, to speak with enthusiasm of the pleasure and profit which he enjoyed from the uninhibited character of that Society's meetings, in comparison with the more formal atmosphere of such meetings in his own country.

Loewi also paid a longer visit to Cambridge. According to my memory he stayed there for a week or two; and, as I have suggested in another connexion, I think that the contacts which he made there and the ideas which he then encountered, probably had a greater influence than he consciously recognized, in prompting his own best-known experimental achievement, nearly two decades later. Apart from the meeting with Hopkins, which I have mentioned already, Loewi had the opportunity, which he would certainly have used to the full, of learning about the several productive lines of research which were then in progress at Cambridge, in the narrow compass of its old Physiological Laboratory. And, of these, the one which he specially remembered was his first acquaintance, at close quarters, with Langley and Anderson's classical investigation of the detailed structure, relations and contrasted functions of the two divisions – sympathetic and parasympathetic in Langley's nomenclature – of the involuntary, or autonomic, nervous system. It had long been known that the actions of certain vegetable alkaloids showed a curiously limited relation to functions of certain of the autonomic nerves, which had now been clearly assigned to the parasympathetic division; the effects of impulses in these being specifically paralyzed by atropine and, on the other hand, reproduced with a mimetic accuracy by muscarine. And in years then still recent, a comparably close, mimetic relationship had been observed by Lewandowsky, and a little later confirmed and extended by Langley himself, between the effects produced by extracts from the medulla of the suprarenal gland, and those of impulses in nerves of the thoracicolumbar, sympathetic division of the autonomic system. T.R. Elliott also, then a young graduate working at Cambridge with the Coutts-Trotter Research Studentship, had already, at the time of Loewi's visit, begun the series of brilliant studies of the actions of autonomic nerves, and especially of the then still puzzling fidelity with which those of the sympathetic division were reproduced by adrenaline – studies which were to lead him, only two years later, to make the first published suggestion, brief but quite definite, of a chemical transmission of the effects of sympathetic nerve-impulses, by the release of adrenaline at the nerve-fibre terminals. I do not suppose that Loewi, at the time of this first visit to Cambridge in 1902, would yet have heard any explicit suggestion of such a possibility; though he certainly met Elliott, and spoke to me especially, on his return to London, of the impression of brilliant promise which he had received from him. As I have already suggested elsewhere, however, speculations about the meaning of these specifically neuromimetic effects were beginning to be current about that period, and especially in Cambridge circles. The late W.E. Dixon had then recently arrived there, to start an experimental course in pharmacology; and, soon after Elliott's publication, in 1904, of his suggestion of a sympathetic transmitter action for adrenaline, Dixon was to suggest a comparable function for muscarine in relation to parasympathetic effects, and even to put on record an attempt – though not a seriously successful, or a well conceived one – to find experimental evidence for it. It seems reasonable, in any case, to suppose that ideas of the kind were waiting, as it were, to emerge in the Cambridge community, during Loewi's visit there in 1902; and when he himself in the following year, after he had returned to Marburg, made an early and independent mention of the idea that muscarine might be released, to transmit the effect of vagus nerve impulses to the heart. But although Loewi, in much later years, was not unnaturally disposed to attribute some retrospective importance to this incident, it seems clear to me that it amounted to nothing more than a casual suggestion, which he made to the late Walter Fletcher whom he had also met in Cambridge, and who, in 1903, was making a short visit to Meyer's Department in Marburg. Loewi, in any case, had entirely forgotten the occurrence; it was Fletcher who reminded him of it, after 1921, in speaking of the important discovery which Loewi had then recently published. And, in any case, a potent and stable poison such as muscarine was a very unlikely candidate for the suggested transmitter function.

1902–1921

This middle period of Loewi's career in research and teaching included the later years at Marburg, till 1904, when he moved with Hans Meyer to Vienna, remaining there till 1909, when he moved again to Graz, on being himself appointed Professor and Head of the Department of Pharmacology in its University. In spite of these two removals, involving changes in personal associations and academic responsibilities, the record of Loewi's research activities shows no conspicuous breach of continuity, or any alteration in the range and variety of his scientific interests. In a brief review of all his experimental undertakings from 1902 till 1921, as shown in his publications over that period, we may take note of the emergence, from time to time, of items of discovery which were eventually to find their places in the chemical transmission scheme; though he and his collaborators do not appear to have had any direct awareness of this significance at the time when the observations were made.

On his return to Marburg, later in 1902, Loewi had resumed an important series of studies on the physiology and pharmacology of the function of the kidneys, which he had apparently begun already before his visit to England. The results of these were to form the subjects of a series of five communications, published from 1902 to 1905, by Loewi alone for the first two, and in joint authorship with visiting collaborators for the other three – Walter Fletcher, whose visit has already been mentioned, N.H. Alcock, who was later to go to Toronto as Professor of Physiology, and V.E. Henderson, who was already Professor of Pharmacology there, and who was to make more than one such co-operative visit to Marburg. There were, indeed, two further papers published by Loewi and Henderson, after the series on renal function, and making, perhaps, a nearer approach in their interest to the chemical transmission theme, though with no conscious reference to it. These dealt respectively with the effect of stimulating vasodilator nerves, and with the influences of pilocarpine and atropine on the rate of circulation through a salivary gland. Also in 1905 came a publication, jointly with Hans Meyer, on a comparative investigation of the activities of synthetic amines related to adrenaline, and especially of 'arterenol' (nor-adrenaline). These had been offered for investigation by the research staff of the Hoechst Farben-Fabrik. Loewi and Meyer compared adrenaline and its primary homologue for their pressor activities, and observed no important difference. H.D. Dakin, working in England, had independently been making a similar series; and he also had observed a similar approximate equivalence, in pressor action, between adrenaline and nor-adrenaline. It was some years later, in 1910, that I myself examined a wider series of amines, for the inhibitor as well as the augmentor components of their 'sympathomimetic' actions, and then observed that nor-adrenaline had, in its action as seen in the cat, a smaller ratio of inhibitor to augmentor effects, than adrenaline had; and that, in this respect, it reproduced the total effects of sympathetic nerve stimulation in that species, with a more complete accuracy than did adrenaline itself. My own short-sighted reaction to this discovery, however, had been to regard it merely as creating a difficulty, of detail for Elliott's suggestion of adrenaline as the sympathetic transmitter. And it was not till many years later, that Bacq, von Euler and others recognized, and eventually proved, that nor-adrenaline, which I had known in 1910 only as a synthetic artifact, was indeed the principal component, in mammals, of the peripheral sympathetic transmitter.

Loewi seems to have been engaged largely with other calls on his time for a few years after 1905; so it was not till 1908 that his name appeared again on publications recording the results of researches, mostly in collaboration with others, and on several widely different subjects. There were publications in that year with Hans Meyer on varying degrees of sensitiveness to tetanus; with Neubaur on phloridzin diabetes again, and the effect of diuretics on it; with Jonescu on a specific action of digitalis on renal function; and by Loewi alone on a new function of the pancreas. In addition to these, there were two with Fröhlich, one of a more general character, on functions of the autonomic nerves and the effects of drugs upon them, and the other on the occurrence of vaso-constrictor fibres in the chorda tympani. Neither of these, however, had more than a distant relation to the chemical-transmission story. Of much greater significance for its eventual development, though this was still unrecognized at the time of their publication, were two communications which followed, in 1909. One of these was by Loewi with Fröhlich, on the increased sensitiveness to adrenaline produced by cocaine, and the other with Mansfeld, on the action of physostigmine, in increasing responses to the stimulation of parasympathetic nerves.

Thenceforward, however, Loewi appears to have distributed most of his research energies and his attention, for some 12 further years, among the problems which have already been mentioned as recurrently attracting his interest. Treatment with digitalis, and the relation between the action of that drug and that of calcium ions, provide the subject of several of his publications in this period. Carbohydrate metabolism and experimental diabetes figure again as the subjects of several papers, and again in relation to the composition in kations of the nutrient medium for the heart, or other tissue, on which the observations are made. There is another series dealing with the action of the vagus on the heart, which again includes the effects, upon this action, of varying the calcium-content of the saline fluid, as well as the effect of other drugs upon the efficiency of the vagus stimulation. And at the very end of this period, in 1921, came three more publications, the fourth, fifth and sixth of the series, on the relation between the actions of the 'heart-drugs' – digitalis, etc. – and those of the physiological kations, especially of the calcium ion.

Early in this major and final period of Loewi's tenure of the Chair at Graz, he achieved and published his first demonstration of the chemical transmission of effects from the nerve-endings of the autonomic system; and during the remainder of it, he and his co-workers published the series of further papers, which reported their developments in detail of this very important discovery. Very naturally it soon won for Loewi a high reputation among medical research workers, widely beyond the circle of those with specially pharmacological interests. For that reason, it seems more suitable to devote a separate section to its closer consideration, and here to make briefer mention of other lines of research which Loewi also followed actively, with different collaborators, during this same final period, until he was compelled to leave Austria.

A glance at the bibliography, prepared from a manuscript left by Loewi himself and reproduced at the end of this Memoir, will be enough to show that some of these other groups of researches, published during this same period, represented renewals of his attack on pharmacological problems into which he, with a varied succession of colleagues and visiting investigators, had already begun his experimental enquiries, during earlier periods of his research activities. Such were, for example, his further renewed studies of the part played by the inorganic ions, and especially by the calcium content, of the blood or of an artificial nutrient fluid, in the effects of the glucosides of the digitalis series on the activities of the heart muscle, as well as in other drug actions. Another of his interests had been in the nature of diabetes, and its modification by drug actions; quite early in his research career, as we have seen, he was using the diabetic condition, as artificially produced by the administration of phloridzin, for such investigations. It is, accordingly, not surprising to find that the first publication of the discovery of insulin by Banting and Best, in 1921, and the immediate and worldwide interest which it excited, should have made Loewi, with many other investigators in many countries, eager to take part in the more fundamental exploration of the nature of the action of a hormone of such importance for normal physiology, as well as for special therapeutics. Accordingly, in spite of the primary claim upon his own research interest by his own demonstration of chemical transmission, published also in 1921, we find him producing, mainly from 1925 to 1928, with his colleague H. Häusler, another series of papers on the results of their experiments concerning the mode of the action of insulin. Other examples could be given; but a mere inspection of the list of his publications, over the forty years or so of his full and unhindered activity, should suffice to make clear the unusual width and variety of Loewi's general enterprise in research. There can be no doubt, indeed, that his experimental work with his own hands, and the inspiration and guidance which he gave to the many distinguished investigators who had been his assistants and students, would have entitled Loewi to high rank among contributors to the progress of scientific knowledge in many departments of pharmacology and biochemistry, even if he had never been prompted to step suddenly beyond the normal limits of the research programme which he seemed thus to have been planning for himself, in order to carry out the simple but convincing experimental test, which immediately gave a new reality and impetus to the then rather widely discredited theory of the chemical transmission of nervous effects. There can, I think, be no doubt that this new adventure by Loewi, into a field of experiment and ideas in which he had till then shown no direct or conscious interest, was responsible for raising his status in the world of scientific opinion, from that of one of a number of distinguished contemporary pharmacologists, to that of a memorable pioneer, whose work opened a new vista of advance, in the special area of physiological and medical knowledge with which it was concerned.

Loewi's Chemical Transmission Discovery

In the 'autobiographic sketch' already mentioned, which Loewi contributed in 1960 to 'Perspectives in biology and medicine', he makes a kind of precautionary disclaimer, about the accuracy to be expected of such personal records. '... Yet fiction may come into play,' he writes, 'because a retrospective report may not always truly reflect the past as it happened.' Anybody, I suppose, who has found himself called upon, over the years, to repeat his account of a personal experience on numerous occasions, must have been conscious of a temptation, at least, to make it more acceptable, by such trimming of details, or shifts of emphasis, as could be effected without weakening its essential veracity. We are entirely dependent, of course, on Loewi's memory, for detailed knowledge of the extraordinary circumstances in which he was led to perform his rightly famous experiment; but, though he certainly had no lack of sprightly imagination, or of gift for picturesque description, I can testify that his own account of this incident has been free from such changes during the four decades since he first described it, except in a minor detail of timing. I suppose that I must have been one of the earliest to hear from him the remarkable story of the dream, from which he woke on two successive nights, with a strong conviction that the experiment which it had suggested was one of great importance. On the first occasion he had seized a scrap of paper, made on it what he thought to be a sufficient record, and fallen asleep again; only to find, next morning, that the scrawl on the paper was illegible, and that he had completely forgotten what it was about. According to my own memory of what he then told me, on waking again from the same dream on the following night, and feeling naturally determined not to lose the clue again, he made a careful record of what it had suggested; so that, on the following morning, he was able to take the chit with him to the laboratory, carry out the proposed experiment and obtain the predicted result. Loewi's own final version of the affair, as given in the 'autobiographic sketch', differs from that which I had thus remembered, only in a detail which, perhaps, gives it an even stronger dramatic appeal. It describes how, on his second waking from the insistent dream, he was so anxious not to lose its message again, that he rose from bed immediately, at 3 a.m., and hurried straight to the laboratory, to perform the successful experiment then and there, in the small hours. Obviously this, his own direct memory of how it happened, has a prior claim to acceptance, over one in which my own fallible memory was also involved.

In any case, however, the difference would have no kind of significance for the cardinal importance of the result of the actual experiment, which demonstrated for the first time the reality of the intervention of a chemical agent, in the transmission of the excitatory effect of a nervous impulse, across the small breach of continuity at neuro-effector or inter-neuronal junctions. So far, a process of this kind had been no more than a speculative suggestion. If its truth could have been experimentally established, it could have been accepted with confidence, to account for certain special features of the transmission at such junctions – the time-delay which it involved and its unidirectional character, for which there had been otherwise no satisfactory explanation; and, more particularly, the suggestively localized and closely neuro-mimetic actions of certain natural bases, especially of adrenaline and some of its homologues and, more recently, of acetylcholine. Hitherto, however, none among those of us who had been taking a particular interest in such speculations, had been able to think of a method by which they could be put to the practical test of experiment. And now Otto Loewi, who, by his own confession, had not paid any conscious attention to such ideas, at least for many years, had found himself mysteriously subjected to the repeated pressure of a suggestion, welling up as it were from some deeper sub-conscious level, and urging him to perform an experiment so simple in its design and its technique, that any reasonably dexterous student could easily have carried it out. And this experiment was actually to give him a simple and unambiguous answer, establishing the chemical transmission of nervous effects, at least at peripheral, autonomic, neuro-effector junctions, as an experimental fact.

The position appears to offer ample scope for psychological speculation. Loewi has himself recorded his own conviction, that, if he had considered the matter carefully in the daytime, he would undoubtedly have rejected the kind of experiment suggested by his dream, on the ground that, if a chemical agent were actually concerned in the transmission, it could not be expected to appear in more than the minute quantities required to act upon the immediately contiguous, effector cells. It may be, indeed, that similar considerations would have deterred even some of those who had retained a more conscious interest in the theory than Loewi had, from using so simple a means of putting it to the test. I do not believe, in fact, that any of them had thought of it. If I had done so myself, I do not think that I should have brushed it aside, as too simple to be worth trying. I find myself more in sympathy with an unidentified German colleague, who happened to be standing behind me when Loewi gave a demonstration of his experiment, at an International Physiological Congress in 1926. Seeing the ease with which the vagus inhibition of one isolated frog's heart could be conveyed to another, by a mere transfer of the saline contents with a pipette, he exclaimed, 'Ach! Warum habe ich nicht daran gedacht?' Why, indeed, should not anybody interested in such matters have thought of it, and tried it, at any time during the previous twenty years? The fact is that nobody did; and the admiration which I felt for Loewi's achievement was only enhanced, by the thought that his relative aloofness from this problem might well have made it easier for him to follow the mysterious prompting of his dream, and thus to adopt so elegantly simple a plan of experiment, to make with it, at one single step, so revolutionary an advance in the position of the chemical transmission theory, and thus to open the way widely to so much that has followed.

There was one feature, however, of Loewis's first published account of this crucial experiment, which I found more difficult to understand. In referring to the two different transmitters which he found to appear in the saline contents of the heart, in response to the stimulation of its inhibitor vagus nerve and of its augmentor-accelerator sympathetic nerve-supply, he referred to them merely in terms of their respective origins, as 'Vagus-stoff' and 'Accelerans-stoff'. He had thus appeared to ignore, or to be unaware of, the fact that there were two good chemical candidates, the respective sympathomimetic and parasympathomimetic activities of which had long entitled them, in theory, to be considered for the transmitter roles, if once the real existence of such a chemical transmitter function could be demonstrated. At our next encounter I ventured to rally him on this reticence. 'Why', I asked him, 'did you not say frankly that your vagus-substance might well prove to be something like acetylcholine?' He had replied that, of course, he believed it to be acetylcholine; but that, having recently, in an entirely different connexion committed himself too rashly to a conclusion which he had subsequently had to withdraw, he was abnormally sensitive to the danger of publishing another speculative suggestion, lest it might prove also to be premature. Most fortunately, however, this attitude, which seemed to me to be one of excessive caution, did not prevent him, in the years that followed, even almost to the time when his research activities were rudely interrupted in 1938, from pursuing the comparison between the properties and behaviour of acetylcholine and those of his 'vagus-substance', by such methods as were applicable to the chemical and pharmacological comparison of extremely minute quantities. Many of these experiments were carried out in association with a succession of collaborators – Navratil, Engelhart, Hellauer. And their results, as a whole, not only succeeded in establishing the identity of the parasympathetic transmitter with acetylcholine, in all the properties and activities for which it proved practicable to compare them, but threw much additional light on the occurrence in nervous tissues of a substance with these characteristics. It was especially significant for further developments, that Loewi and his co-workers found that the vagus-transmitter, when liberated, was extremely susceptible to destruction by an enzyme, which was not merely, as I indeed had much earlier suggested, an esterase, but a specific cholinesterase, and that the action of this was readily inhibited by the alkaloid eserine (physostigmine), the specific pupilloconstrictor action of which had been shown as long ago as 1905, by H.K. Anderson at Cambridge, to be due to its intensification and prolongation of the effects of impulses in the (parasympathetic) oculi-motor nerve-supply to the pupil. This rationalisation of the effects of eserine, as essentially due to its prevention or delay of the destruction of acetylcholine, by its antagonism to the action of a cholinesterase, was to be of outstanding importance for the further study of cholinergic, chemical-transmitter mechanisms, and especially so for the extension of such experiments to the organs of warm-blooded vertebrates, and to those activities of acetylcholine which I had described as its 'nicotine' actions; the studies of Loewi and his colleagues having been limited, at least in intention, to its peripheral, 'muscarine' actions. Loewi, indeed, seems to have adopted, for a time, an attitude of almost obstinate skepticism, to what others of us had regarded as a possibility at least to be tested; namely, that these 'nicotine' actions of acetylcholine, on ganglion cells and on the motor end-plates of voluntary muscle fibres, raised the question whether even the transmission of the rapid effects of nerve impulses across such synaptic junctions, might not also be effected by the liberation there of acetylcholine. In the first Harvey Lecture which Loewi gave by invitation in New York, in 1933, he even, as it seemed to me, went out of his way to give public expression to his disbelief in such a chemical transmission of the effects of voluntary nerve impulses to voluntary muscle fibres. He was unaware, of course, that, at the time when he was thus disclaiming belief in such a possibility, some of my own colleagues were busy already with plans for putting it to the test of direct experiment, an essential condition of which would be the use of the protective action of eserine, which Loewi and his colleagues had discovered. By the time that he gave the Ferrier Lecture to the Royal Society, in 1935, on what he still preferred to call the 'humoral' transmission of nervous impulses, the experimental evidence for such a process in the excitation of voluntary muscle fibres from motor nerve endings, and for acetylcholine as again the agent of the chemical transmission, had already been published in sufficient detail to carry conviction, and with Otto Loewi as with other experts in such matters. He was now able, accordingly, to give public acceptance to the identification of the transmitter from peripheral parasympathetic nerve-endings, at ganglionic synapses and from voluntary motor nerve endings, as acetylcholine itself; though he still preferred – rather surprisingly, I thought – to reserve judgement, with regard to the presumptive identification of the peripheral sympathetic transmitter, his own 'Accelerans-stoff', as adrenaline (or its primary homologue), although the first published proposal of the transmitter conception had been made by Elliott, more than thirty years earlier, to explain the closely sympathomimetic action of adrenaline, on effector cells which were, or had been, sympathetically innervated.

During the passage of more than a quarter of a century, the development of knowledge concerning these transmitter mechanisms has, of course, proceeded very far beyond the point which it had reached in 1935, when Loewi gave his Ferrier Lecture. The chemical identification of the agents of cholinergic and adrenergic transmission has long since passed beyond the stage of doubt, or controversy; while, through the brilliant experimental achievements of Sir John Eccles and many others, the occurrence of a chemical transmission process at most of the synaptic junctions in the vast neuronal complexities of the mammalian central nervous system, seems already to have been firmly established in principle; although it seems further clear that other chemical transmitters, than those now accepted as functional in the peripheral system, have yet to be recognized and identified as responsible for the transmitter functions at central synapses. Such developments, and others, perhaps, not yet foreseen, are leading far from any which could have been contemplated when Loewi, in 1921, performed his simple but convincing demonstration with the isolated hearts of frogs; but it was this which first brought the chemical transmitter conception within the range of directly experimental proof and investigation, and opened the way to all that has followed, in that still expanding field of enquiry.

The award of a Nobel Prize to Otto Loewi for his part in these researches, in 1936, was greeted with general acclaim; and to me it was a matter for special pride and satisfaction to be joined with him in that award. By both of us it was greeted as a welcome new link in a scientific and personal friendship, already of long standing.

Personal and Family Life. Exile and Re-Settlement

Mention has been made already of the unusual width and variety of Loewi's cultural interests, even while he was still at the student stage of his university career. Throughout his life he seems to have retained this eagerness for new artistic and cultural experiences and contacts. He himself records that, having been naturally endowed with no special aptitude for making or listening to music, he set himself deliberately, and successfully, to acquire an appreciation of its intellectual and emotional appeal, with a resulting enthusiasm which came to take high rank in the scale of his pleasures. His enthusiasms and his dislikes were easily aroused, and given prompt and effective expression. He had an unabashed enjoyment of good food, good wine and good conversation, being himself a ready and uninhibited talker, but always worth hearing and never in danger of being a bore. He had a genuine enthusiasm for scientific ideas and achievement in general, as well as for those in his chosen lines of research activity. But, in addition, he had, in an eminent degree, the characteristics best summarized in his native language; he was a genuine 'Lebenskünstler'. With it all, he was capable of giving, and attracting, consistent and loyal friendship.

Loewi had found Marburg, where he held his first academic appointment with Hans Meyer, relatively poor in its social and cultural opportunities; but these could nowhere have been more lavishly at his disposal than in Vienna, whither he moved with Meyer in 1904. During his stay there he formed the habit of spending his annual holiday in Switzerland, and acquired a lasting affection for the Engadine in particular. It was in Pontresina, in 1907, that he first met Professor Guido Goldschmiedt, then Professor of Chemistry in Prague and Vienna, with his wife and their daughter Guida, who became Loewi's wife in 1908, and shared with him all the details of the removal to Graz in 1909, and of the establishment there of the attractive new home, where they were to spend the next twenty-nine happy years. Guida Loewi was to become the able mistress of their charming household, mother of their family of three sons and one daughter, perfect hostess to their many friends, constant and devoted companion to her husband and sharer in all his interests and experiences, happy, tragic and adventurous, until she died with a tragic suddenness in 1958, fifty years after their marriage. They had by then been living for some nineteen years in New York, where friendly invitation had enabled them to become peacefully re-settled with their family, after their expulsion from Austria twenty years earlier.

Till the final blow came, Graz, the second city of Austria, with its second Austrian university and medical school, had offered Loewi what seemed to be ideal conditions for his academic activities of teaching and research in his own chosen subject, together with the social and cultural privileges which meant so much to him and his wife and their growing family – drama, opera, concerts of music, museums of art and antiquity; not, of course, on the scale of the historic and world-famous institutions of Vienna, but excellent in their own setting, which, on the other hand, was even desirably free from the political and other distractions of what was still the Imperial Capital of a large part of Eastern Europe, when Loewi left it for the relative seclusion of Graz, with its attractive landscape setting. Even the First World War, beginning five years later, seemed to leave him, in his Graz Professorship, enviably free, by comparison with his colleagues in the other belligerent countries, to pursue the even tenor of his academic researches, little hampered by the demands of war-time emergencies.

After all these happy and fruitful years, amounting to about three-quarters of the whole period of Loewi's full scientific activity, and when the time for his normal retirement must already have been coming into view, the blow fell suddenly, in March 1938. With the Nazi occupation of Austria, German troops marched into Graz. The eldest son of the Loewis, and their daughter, both had the good fortune to be outside Austria at the time; but Loewi himself and the two younger sons were arrested and thrown into prison, together with a large number of other Jewish residents in Graz. He was liberated after two months of this brutal deprivation, and his sons a month later; but it was September of that year before he obtained permission to leave Austria, and then only at the cost of signing an instruction to a Swedish bank, to transfer securities, which were being held there for him, to a German bank under Nazi control. His cable, informing me that he was flying to London, found me urgently engaged with measures for safeguarding the valuable items of equipment and library in the National Institute for Medical Research, against risk of damage in the war, which all England was even then expecting to begin. That immediate danger, however, was deferred for a year, and Loewi arrived safely and in remarkably good form. I have never ceased to admire the dauntless courage and enterprise with which he, with many others in comparable predicaments at that time, refused to waste time and energy in merely futile resentment, but devoted their abilities promptly to the quest for new careers, to the scientific advantages of the countries which offered them refuge and new opportunities. Loewi stayed with us then for a few weeks, during which he took me to call on Sigmund Freud, whom he had known in Vienna, and who was then resident in Hampstead, but already sadly afflicted by the jaw cancer, from which he was to die. Loewi then accepted a temporary appointment under the Belgian Foundation Franqui, and left for Brussels. Returning to friends in England for a holiday in the next year, 1939, he was caught by the outbreak of the Second World War, and found harbourage and opportunity in the Pharmacology Department under the late Professor J.A. Gunn at Oxford, until the middle of 1940. Then he was able finally to accept an invitation from the Medical School of the New York University, to join their faculty as a Research Professor of Pharmacology, with accommodation in the Department of Professor George B. Wallace, whose initiative was doubtless concerned in the proposal. Loewi arrived in New York on 1 June 1940, to take up this appointment, and thus to settle himself into a further and final stage of his career, at the age of 67. His wife and the other members of his family were able to join him in New York during the following year. Their financial means, in relation to the costs of living, were not, of course, to have the comfortable amplitude which they had enjoyed in Graz; but they readily and gratefully adapted themselves to more frugal standards. He became an American citizen in 1946.

Of the many honorary Degrees which Loewi received in these later years from universities in different countries, including those of Graz and Frankfurt, and of the honorary memberships to which he was elected, none gave him greater pleasure than his election to the Foreign Membership of The Royal Society in 1954.

This happy evening to Otto Loewi's scientific career, which American generosity and respect for his achievement had provided, was to last for rather more than another twenty-one years. It was not to be expected that his work in the laboratory would produce further outstanding discoveries at that stage, and under these new conditions. He was able to tidy a few 'loose ends' of the work which had been so brutally interrupted, to regain his poise, and, as the years passed, to spend more of his time in writing articles and giving lectures, for which American centres of medical research were characteristically and increasingly generous with requests and invitations, as the rumour of what he had to offer grew and spread more widely. He and his wife were able, quite early, to form the habit of going annually, when the summer heat of New York became oppressive, to Woods Hole on the Massachusetts coast, with its world-famous centre for researches in marine biology. There, for the rest of his life, he was able each summer to acquire new scientific interests, to make new friendships, and to renew others which had already been formed. As already mentioned, the long life-partnership was sadly terminated by Mrs Loewi's sudden death in 1958; and, not long afterwards, Loewi himself was victim of a serious accident, when about to leave hospital. He had a fall, which broke his pelvis and seriously disabled him, and from the physical effects of which he never fully recovered. When I last saw him in New York, in October 1959, he could just push himself about in his apartment, with the support of a strong metal frame; but his brain was as alert as ever, as eager for reminiscences and discussion, and for new mental enterprise. Nor was he to be deprived of his annual enjoyment of months at Woods Hole; for

American friends rallied round him in each of the remaining summers, and transported him thither by car. In the last letter which I received from him, dated from Woods Hole on 2 July 1961, he wrote: 'I love this place more than any other in the world known to me, except the Engadine in Switzerland.' It was good to think that he had thus been able, at last, to enjoy the two places which, in his affectionate judgement, were the best of both worlds.

And when Otto Loewi's life reached its natural ending on Christmas Day 1961, in his 89th year, its circumstances were such as he himself, with a connoisseur's instinct, might well have approved, or even desired, if he could have foreseen them. He received, on Christmas Eve, from a friend – maybe from Woods Hole – the present of a fine lobster. This being much to his liking, he ate the whole of it for his dinner, with a bottle of wine, which he shared with his nurse-housekeeper. And then, on the following morning, when he was speaking to a friend with his usual eager impetus, he stopped in the middle of a sentence; and they found that his life had flickered out.

The Director of the Woods Hole Laboratory, and his associates, with others who had there learned to know and to admire Otto Loewi, arranged for the re-interment of his ashes in the Churchyard there, on 17 August 1962, and for a commemorative lecture and meeting, at which his friends could recall his personality and the sparkle of his mind, and all that had so greatly endeared him to their community. I myself was unfortunately prevented, by a minor accident, from accepting a generous invitation to be present on this occasion, and to take part in the proceedings. Fortunately my friend and former colleague, Sir Lindor Brown, Sec. R.S., was able to be present, and to represent many British friends and admirers of Otto Loewi.

I have been greatly indebted to Otto Loewi's youngest son (now domiciled in England as Geoffrey Low) for information about his last hours, and for access to the documents which he left, including the MS. Bibliography, in the reading and verification of which I have been greatly assisted by Miss M.A.B. Venables, of the Library Staff of the Royal Society of Medicine. In the same connexion I have received greatly valued assistance from Professor Otto Krayer of the Harvard Medical School, and from the Springer-Verlag of Heidelberg. For the portrait here published I have been greatly indebted to Lady Simon, who kindly lent me a print from a photograph taken by her husband, the late Sir Francis Simon, F.R.S., when Otto Loewi was in Oxford in 1940. It seems to me the most characteristic portrait of him which I have seen, and shows him at the age of 67, between his leaving Austria and his migration to America. For the preparation, from the original small print, of an enlargement more suitable for reproduction, I am indebted to Mr C.E. Fryer of the Art Department of the Wellcome Foundation Ltd.

Bibliography

ARTICLES IN BOOKS, ETC.

1900 *Untersuchungen über den Nucleinstoffwechsel.* Habilitationsschrift, Marburg.

1907 Arzneimittel und Gifte in ihrem Einfluss auf den Stoffwechsel. In: Von Noorden. *Handbuch der Pathologie des Stoffwechsels.* 2. Aufl. 2, 663–831. Berlin.

1907 Also trans.: Drugs and poisons – their influence on metabolism. In: Von Noorden, *Metabolism and practical medicine*, English issue. 3, 1061–1198. London.

1922 Allgemeines über Beziehungen zwischen Nervensystem und vegetativen Funktionen. In: Lüdke-Schlayer, *Lehrbuch der pathologischen Physiologie*, pp. 258–268. Leipzig.

1937 Über postganglionäre Gefässwirkungen von Nicotin and Acetylcholin. *Arch. int. Physiol., Vol. jubilaire pour J. Demoor*, p. 349.

1950 Vistas in pharmacology. *Symposium on the next half century in medicine.* Starling-Winthrop Research Institute, pp. 13–24. Rensselaer, N.Y.

1952 On the mechanism of drug action. In: Barron, E.S.G. (ed) *Modern trends in physiology and biochemistry.* pp. 405–425. New York.

1953 *From the workshop of discoveries. Porter Lectures, Ser. 19.*
(1) Reflections on the study of medicine.
(2) From the workshop of discoveries.
(3) Problems in the field of adrenal function.
Lawrence, Kan: Univ. Kansas Press.

PAPERS IN JOURNALS

1896 Zur quantitativen Wirkung von Blausäure, Arsen und Phosphor auf das isolirte Froschherz. *Arch. exp. Path. Pharmak.* 38, 127–138.

1898 Über das harnstoffbildende Ferment der Leber. *Hoppe-Seyl. Z. physiol. Chem.* 25, 511–522.

1900 Beiträge zur Kenntniss des Nucleinstoffwechsels. 1. Mitteilung. *Arch. exp. Path. Pharmak.* 44, 1–23.

1901 Zur Kenntnis des Nucleinstoffwechsels. *S.B. Ges. Naturw.* Marburg (1900), 1901, 89–92.

1901 Pharmacodynamische Untersuchungen über Anagyrin. *Arch. int. Pharmacodyn.* 8, 65–75.

1901 Zur Lehre der Fettresorption. *S.B. Ges. Naturw. Marburg*, 1901, 90.

1901 Untersuchungen über den Nucleinstoffwechsel. (II. Mitteilung.) *Arch. exp. Path. Pharmak.* 45, 157–185.

1901 Über die Stellung der Purinkörper im menschlichen Stoffwechsel. Bemerkungen zu der gleichnamigen Untersuchung von Burian und Schur. *Pflügers Arch. ges. Physiol.* 88, 296–298.

1902–2 Zur Frage nach der Bildung von Zucker aus Fett. *Arch. exp. Path. Pharmak.* 47, 68–76.

1901–2 Über den Einfluss des Camphers auf die Grösse der Zuckerausscheidung im Phlorhizindiabetes. *Arch. exp. Path. Pharmak.* 47, 56–67.

1901–2 Zur Kenntniss des Phlorhizindiabetes. *Arch. exp. Path. Pharmak.* 47, 48–55.

1902 Untersuchungen zur Physiologie und Pharmakologie der Nierenfunction. *Arch. exp. Path. Pharmak.* 48, 410–438.

1902 Über Eiweisssynthese im Tierkörper. *Arch. exp. Path. Pharmak.* 48, 303–330; Also – *Zbl. Physiol.* 15, 590–591.

1903 Untersuchungen zur Physiologie und Pharmakologie der Nierenfunktion. II. Mitteilung. Über das Wesen der Phlorhizindiurese. *Arch. exp. Path. Pharmak.* 50, 326–331.

1904 Pharmakologie des Wärmehaushalts. *Ergebn. Physiol.* 3, 332–372.

1904 Über den Diastasengehalt verschiedener Blutsera. *S.B. Ges. Naturw. Marburg,* 100–102.

1905 Zur Kenntnis der Tetanusvergiftung. *S.B. Ges. Naturw. Marburg, 11–13.*

1905 Ueber das Wesen der Coffeindiurese. *S.B. Ges. Naturw. Marburg, 76–79.*

1905 Zur Physiologie und Pharmacologie der Vasodilatatorenreizung. *S.B. Ges. Naturw. Marburg*, 79.
1905 (With W.M. FLETCHER and V.E. HENDERSON.) Untersuchungen zur Physiologie und Pharmakologie der Nierenfunction. 111. Mitteilung. Über den Mechanismus der Coffeindiurese. *Arch. exp. Path. Pharmak.* 53, 15–32.
1905 (With N.H. ALCOCK.) Untersuchungen zur Physiologie und Pharmakologie der Nierenfunction. IV. Mitteilung. Über den Mechanismus der Salzdiurese. *Arch. exp. Path. Pharmak.* 53, 33–48.
1905 (With V.E. HENDERSON.) Untersuchungen zur Physiologie und Pharmakologie der Nierenfunction. V. Mitteilung. Über den Mechanismus der Harnstoffdiurese. *Arch. exp. Path. Pharmak.* 53, 49–55.
1905 (With V.E. HENDERSON.) Über die Wirkung der Vasodilatatorenreizung. *Arch. exp. Path. Pharmak.* 53, 56–61.
1905 (With V.E. HENDERSON.) Über den Einfluss von Pilocarpin und Atropin auf die Durchblutung der Unterkieferspeicheldrüse. Ein Beitrag zur Frage nach den Beziehungen zwischen Organtätigkeit und Organdurchblutung. *Arch. exp. Path. Pharmak.* 53, 62–75.
1905 (With H.H. MEYER.) Ueber die Wirkung synthetischer, dem Adrenalin verwandter Stoffe. *Arch exp. Path. Pharmak.* 53, 213–226.
1905 (With T. ISHIZAKA.) Über die Wirkung von Muskarin auf das nicht oder unzureichend gespeiste Froschherz und die Gegenwirkung von Kalziumsalz. *Zbl. Physiol.* 19, 593–595.
1906 (With A. FRÖHLICH.) Ueber vasokonstriktorische Fasern in der Chorda tympani. *Zbl. Physiol.* 20, 229–232.
1907 Ueber Wirkungsweise und Indikation einiger diuretisch wirkender Mittel. *Wien. klin. Wschr.* 20, 1–5.
1907 (With A. FRÖHLICH.) Scheinbare Speisung der Nervenfaser mit mechanischer Erregbarkeit seitens ihrer Nervenzelle. (Nach Versuchen an Eledone moschata). *Zbl. Physiol.* 21, 273–276.
1908 (With A. FRÖHLICH.) Untersuchungen zur Physiologie und Pharmakologie des autonomen Nervensystems. I. Mitteilung über die Wirkung der Nitrite und des Atropins. *Arch. exp. Path. Pharmak.* 59, 34–56.
1908 (With E. NEUBAUER.) Über Phlorhizindiurese und über die Beeinflussung der Phlorhizinzuckerausscheidung durch Diuretica. *Arch. exp. Path. Pharmak.* 59, 57–63.
1908 (With A. FRÖHLICH.) Über vasoconstrictorische Fasern in der chorda tympani. *Arch. exp. Path. Pharmak.* 59, 64–70.
1908 (With D. JONESCU.) Über eine spezifische Nierenwirkung der Digitaliskörper. *Arch. exp. Path. Pharmak.* 59, 71–82.
1908 Über eine neue Funktion des Pankreas und ihre Beziehung zum Diabetes mellitus. *Arch. exp. Path. Pharmak.* 59, 83–94.
1908 (With H.H. MEYER.) Über Tetanusgift-Empfindlichkeit und Überempfindlichkeit. *Arch. exp. Path. Pharmak.* Suppl. Bd, pp. 355–365.
1910 (With A. FRÖHLICH.) Über eine Steigerung der Adrenalinempfindlichkeit durch Cocain. *Arch. exp. Path. Pharmak.* 62, 159–169.
1911 (With G. MANSFELD.) Über den Wirkungsmodus des Physostigmins. *Arch. exp. Path. Pharmak.* 62, 180–185.
1911 Pharmakologie und Klinik. *Wien klin. Wschr.* 23, 273–278.
1911 Über Digitalistherapie. *Mitt. Ver. Aerzte Steiermark*, 48, 379–387.
1912 Untersuchungen zur Physiologie und Pharmakologie des Herzvagus. I. Mitteilung: Über den Einfluss von Chloralhydrat auf den Erfolg der Vagusreizung. II. Mitteilung. Über die Bedeutung des Calciums für die Vaguswirkung. III. Mitteilung. Vaguserregbarkeit und Vagusgifte. *Arch. exp. Path. Pharmak.* 70, 323–342; 343–350; 351–368.
1913 Über innere Sekretion. *Mitt. Ver. Aerzte Steiermark*, 50, 71–83.
1913 (With O. WESELKO.) Über die Abhängigkeit experimentell-diabetischer Störungen von der Kationenmischung. Vorläufige Mitteilung. *Münch. Med. Wschr.* 60, 690.
1914 (With O. WESELKO.) Über den Einfluss der Thyreoidektomie auf die Wärmestichreaktion bei Kaninchen. *Zbl. Physiol.* 28, 197.

1914 Über die Folgen der Nebennierenexstirpation beim Frosch. (Dtsch. physiol. Ges., 2.–5. Juni, 1914.) Abstr.: *Zbl. Physiol.* 28, 727.

1914 (With W. GETTWERT.) Über die Folgen der Nebennierenexstirpation. I. Mitteilung: Untersuchungen am Kaltblüter. *Pflügers Arch. ges. Physiol.* 158, 29–40.

1914 (With O. WESELKO.) Über den Kohlehydratumsatz des isolierten Herzens normaler und diabetischer Tiere. *Pflügers Arch. ges. Physiol.* 158, 155–188.

1917 Über Calciumwirkung und Calciumtherapie. *Mitt. Ver. Aerzte Steiermark,* 54, 53–59, 63–71.

1917 Über die Entstehung von functionellen Anpassungen im individuellen Dasein. *Naturwissenschaften,* 5, 501.

1917 Über den Zusammenhang von Digitalis- und Kalziumwirkung. *Münch. med. Wschr.* 64, 1003–1004.

1917–8 Über den Zusammenhang zwischen Digitalis- und Kalziumwirkung. *Arch. exp. Path. Pharmak.* 32, 131–158.

1918 Über den Zusammenhang zwischen Digitalis- und Kalziumwirkung. III. Mitteilung. *Arch. exp. Path. Pharmak.* 83, 366-380.

1918 Über Spontanerholung des Froschherzens bei unzureichender Kationenspeisung. II. Mitteilung: Ein Beitrag zur Wirkung der Alkalien aufs Herz. *Pflügers Arch. ges. Physiol.* 170, 677–695.

1918 Zur Frage der Verwertbarkeit der Glukose bei Diabetes. Ther. *Mh.* 32, 350–354.

1919 (With H. LEIB.) Über Spontanerholung des Froschherzens bei unzureichender Kationenspeisung. III. Mitteilung: Quantitative mikroanalytische Untersuchungen über die Ursache der Calciumabgabe von seiten des Herzens. *Pflügers Arch. ges. Physiol.* 173, 152–157.

1919 Über Wertung und Wirkung von Werken der bildenden Kunst. *Z. Asthet. Kunstwiss.* 14, 239–252.

1921 Über die Beziehungen zwischen Herzmittel- und physiologischer Kationenwirkung. IV. Mitteilung. Über Nichtelektrolytwirkung aufs Herz. V. Mitteilung. Über die Wirkung von Lipoiden auf die Hypodynamie und deren Beziehung zum Kalium. *Pflügers Arch. ges. Physiol.* 187, 105–122; 123–131.

1921 Über die Beziehungen zwischen Herzmittel- und physiologischer Kationenwirkung. VI. Mitteilung: Über die Kaliumcontractur. *Pflügers Arch. ges. Physiol.* 188, 87–97.

1921 Über humorale Übertragbarkeit der Herznervenwirkung. I. Mitteilung. *Pflügers Arch. ges. Physiol.* 189, 239–242.

1922 Über humorale Übertragbarkeit der Herznervenwirkumg. II. Mitteilung. *Pflügers Arch. ges. Physiol.* 193, 201–213.

1922 Weitere Untersuchungen über humorale Übertragbarkeit der Herznervenwirkungen. *Klin. Wschr.* 1, 22–23.

1922 Über humorale Übertragbarkeit der Herznervenwirkung. Naturwissenschaften. 10, 52–55. Also. *Verh. dtsch. Pharmak. Ges. Nr. 1,* xxiv. In *Arch. exp. Path. Pharmak.* 92.

1922 Sulla trasmissibilita umorale dell'azione dei nervi cardiaci. *Biochem. Terap. sper.* 9, 175–191.

1922 (With E. GEIGER.) Über Änderung des Cholingehaltes der Froschmuskulatur durch elektrische Reizung. *Biochem. Z.* 127, 174–180.

1922 (With E. GEIGER.) Versuche über die Glucosepermeabilität der Leber. *Klin. Wschr.* 1, 1210.

1922 (With J. SOLTI.) Über die Wirkung von Pilocarpin und Atropin auf den isolierten Krötenmuskel und ihre Abhängigkeit von der Ionen-Mischung. *Klin. Wschr.* 1, 2046.

1923 Die Ursache der Unempfindlichkeit des Krötenherzens für die Contracturwirkung der Digitalisstoffe. *Pflügers Arch. ges. Physiol.* 198, 359–366.

1923 (With E. GEIGER.) Versuche über die Glucosepermeabilität der Leber. *Pflügers Arch. ges. Physiol.* 198, 633–643.

1923 Über Steuerungen von Funktionen im Tierkörper. *Naturwissenschaften.* 11, 117–123.

1923 (With J. SOLTI.) Über die Wirkung von Pilocarpin und Atropin auf den quergestreiften Muskel. *Arch. exp. Path. Pharmak.* 97, 272–284.

1923 Über die humorale Übertragbarkeit der Herznervenreizungswirkung. Bemerkungen zu dem Aufsatz von H.J. Hamburger in Jg. 2, Nr. 28 dieser Wochenschrift. *Klin. Wschr.* 2, 1840–1841.
1924 (With G. SINGER.) Über die Wirkung des Jods auf die Atmung isolierter Zellen. *Wien. med. Wschr.* 74, 328–332.
1924 Weiteres über humorale Übertragbarkeit der Herznervenwirkung. *Klin. Wschr.* 3, 680–681.
1924 Über humorale Übertragbarkeit der negativ chronotropen und negativ dromotropen Vaguswirkung. *Klin. Wschr.* 3, 1078.
1924 Über humorale Übertragbarkeit der Herznervenwirkung. III. Mitteilung. *Pflügers Arch. ges. Physiol.* 203, 408–412.
1924 Über humorale Übertragbarkeit der Herznervenwirkung. IV. Mitteilung. *Pflügers Arch. ges. Physiol.* 204, 361–367.
1924 Über humorale Übertragbarkeit der Herznervenwirkung. V. Mitteilung: Die Übertragbarkeit der negativ chrono- und dromotropen Vaguswirkung. *Pflügers Arch. ges. Physiol.* 204, 629–640.
1924 (With E. NAVRATIL.) Über humorale Übertragbarkeit der Herznervenwirkung. VI. Mitteilung: Der Angriffspunkt des Atropins. *Pflügers Arch. ges. Physiol.* 206, 123–134.
1924 (With E. NAVRATIL.) Über humorale Übertragbarkeit der Herznervenwirkung. VII. Mitteilung. *Pflügers Arch. ges. Physiol.* 206, 135–140.
1925 (With H. HÄUSLER.) Insulin und die Zuckerverteilung zwischen flüssigen und nicht flüssigen Systemen. (Kurze Mitteilung.) *Biochem. Z.* 156, 295–299.
1925 (With H. HÄUSLER.) Zur Frage der Wirkungsweise des Insulins. I. Mitteilung: Insulin und die Glucoseverteilung zwischen flüssigen und nichtflüssigen Systemen. *Pflügers Arch. ges. Physiol.* 210, 238–279.
1926 (With S. DIETRICH and H. HÄUSLER.) Weiteres über Insulinwirkung und Diabetes. *Klin. Wschr.* 5, 414.
1926 (With E. NAVRATIL.) Über das Schicksal des Vagusstoffs und des Acetylcholins im Herzen. *Klin. Wschr.* 5, 894.
1926 (With E. NAVRATIL.) Über den Mechanismus der Vaguswirkung von Physostigmin und Ergotamin. *Klin. Wschr.* 5, 894–895.
1926 Kritische Bemerkungen zu L. Ashers Mitteilungen. *Pflügers Arch. ges. Physiol.* 212, 695–706.
1926 (With H. HÄUSLER.) Untersuchungen über Diabetes und Insulinwirkung. V. Mitteilung. *Pflügers Arch. ges. Physiol.* 214, 370–379.
1926 (With B. NAVRATIL.) Über humorale Übertragbarkeit der Herznervenwirkung. X. Mitteilung: Über das Schicksal des Vagusstoffs. *Pflügers Arch. ges. Physiol.* 214, 678–688; XI. Mitteilung: Über den Mechanismus der Vaguswirkung von Physostigmin und Ergotamin. *Pflügers Arch. ges. Physiol.* 214, 689–696.
1926 Über die Wirkung des Insulins und des Insulin-Antagonisten des diabetischen Blutes. Nach Versuchen mit S. Dietrich und H. Häusler. *Wien. klin. Wschr.* 39, 1074–1076.
1926 (With S. DIETRICH.) Untersuchungen über Diabetes und Insulinwirkung. VII. Mitteilung. *Pflügers Arch. ges. Physiol.* 215, 78–94.
1926 I. Über das Schicksal des Vagusstoffes im Herzen. II. Über den Mechanismus der Vaguswirkung von Eserin und Ergotamin. (Abstracts of Communications to the XIIth International Physiological Congress held at Stockholm, August 3–6, 1926). *Skand. Arch. Physiol.* 49, 172.
1926 Über den Vagusstoff. *Naturwissenschaften,* 14, 994–995.
1927 Über Strukturfixierung der Glucose und ihre Bedeutung für das Glucoseschicksal. *Naturwissenschaften.* 15, 93–94.
1927 (With M.L. PATRIZI.) Intorno al sopposto 'ormone vagale' di Loewi. (Nuovi esperimenti sui cuori coniugati dei cani di grossa taglia e nuovi argomenti.) *Bull. Sci. med.* 5, 25–43.
1927 (With H. HÄUSLER.) Über die alimentäre Hyperglykämie. *Klin. Wschr.* 6, 313.
1927 (With S. DIETRICH.) Über den Phlorhizindiabetes. *Klin. Wschr.* 6, 313.

1927 (With S. DIETRICH and H. HÄUSLER.) Weiteres über den insulinantagonistischen Stoff. *Klin Wschr.* 6, 856.

1927 (With H. HÄUSLER.) Weiteres über alimentäre Hyperglykämie. *Klin. Wschr.* 6, 856.

1927 (With H. HÄUSLER.) Untersuchungen über Diabetes und Insulinwirkung. IX. Mitteilung: Über das Aufreten des insulinantagonistischen Stoffes im Blut nach Pankreasexstirpation. X. Mitteilung: Weitere Wirkungen des insulinantagonistischen Hormones 'Glykämin' und ihre Bedeutung für den Mechanismus des Diabetes. *Arch. exp. Path. Pharmak.* 123, 56–62; 63–71.

1927 (With H. HÄUSLER.) Über hormonale Vorgänge nach Glucosezufuhr. I. Mitteilung: Über Insulinsekretion nach subcutaner Glucosezufuhr, ihre quantitative Verfolgung und ihren Mechanismus. II. Mitteilung: Über Insulin- und Glykäminsekretion nach peroraler Glucosezufuhr. III. Mitteilung: Versuche am Hungertier. *Arch. exp. Path. Pharmak.* 123, 72–87; 88–119; 120–128.

1927–8 Glykämin und Insulin. I. *Klin. Wschr.* 6, 2169–2176; G Zuelzer: Bemerkungen ... *Klin. Wschr.* 7, 312; O. Leowi – Erwiderung. *Klin. Wschr.* 7, 312.

1928 Zur Frage nach dem Mechanismus der Glykogenolysesteigerung beim Diabetes. *Klin. Wschr.* 7, 311–312.

1928 (With S. DIETRICH.) Insulin und Glykämin. 11. *Klin. Wschr.* 7, 629–634.

1929 Insulin und Glykämin. III. *Klin. Wschr.* 8, 391–393.

1929 (With H. HÄUSLER.) Über die Glucosefixation durch Blutkörperchen. Bemerkungen zur gleichnamigen Arbeit von Högler, Thomann und Überrack. *Biochem. Z.* 214, 229–230.

1929 Bemerkungen zur Receptorenfunktion im Verdauungskanal. *Dtsch. med. Wschr.* 55, 1751–1752.

1930 Hermann Pfeiffer. *Klin. Wschr.* 9, 575.

1930 (With E. ENGELHART.) Fermentative Acetylcholinspaltung im Blut und ihre Hemmung durch Physostigmin. *Arch. exp. Path. Pharmak.* 150, 1–13.

1930 (With E. ENGELHART.) Über die Vagusstoffzerstörung im Herzen. *Arch. int. Pharmacodyn.* 38, 287–291.

1931 Ueber 'neuro-humorale' Auslösungen im Organismus. *Jahreskurse ärztl. Fortbild.* 22, Heft 1, 1–11; also: *Karlsbad. ärztl. Vortr.* 12, 325–342.

1931 Der fehlende Zusammenhang zwischen Herznervenreizung und Bildung eines humoralen Vagusstoffes. Bemerkungen zur Mitteilung von L. Asher und N. Scheinfinkel. *Klin. Wschr.* 10, 1312.

1932 Eröffnungsansprache. *Verh. dtsch. Ges. inn. Med.* 44, 13–16.

1932 Eröffnungsansprache. *Verh. dtsch. pharmak. Ges.* 17–20.

1933 (With E. PICHLER.) Über den Glykogenstoffwechsel des Muskels und seine nervöse Beeinflussung. III. Mitteilung: Ein propriozeptiver glykogenolytischer Reflex. *Pflügers Arch. ges. Physiol.* 233, 51–56.

1934–5 (With M. FLUCH and H. GREINER.) Hypophysenvorderlappen und Glykogenolyse. *Klin. Wschr.* 13, 883–884; also: *Arch. exp. Path. Pharmak.* 177, 167–176.

1935 The Ferrier Lecture on problems connected with the principle of humoral transmission of nervous impulses. *Proc. Roy. Soc. B*, 118, 299–316.

1936 Quantitative und qualitative Untersuchungen über den Sympathicusstoff. Zugleich XIV. Mitteilung über humorale Übertragbarkeit der Herznervenwirkung. *Pflügers. Arch. ges. Physiol.* 237, 504–514.

1937 Über den Adrenalingehalt des Säugerherzens. *Arch. int. Pharmacodyn.* 57, 139–140.

1937 Die chemische Uebertragung der Nervwirkung. *Schweiz. med. Wschr.* 67, 850–855.

1937 (With R. HAGEN, H. KOHN and G. SINGER.) Über eine wasserunlösliche Zustandsform des Acetylcholins im Zentralnervensystem des Frosches. *Pflügers Arch. ges. Physiol.* 239, 430–439.

1938 Über den Mechanismus der Stoffwechselwirkung des Hypophysenvorderlappens. (Pavlov Memorial.) *Fiziol. Zh. SSSR.* 24, 241–244; abstr.: *KongrZbl. ges. inn. Med.* 97, 239.

1938 (With H. HELLAUER.) Vergleich der Acetylcholinausbeute aus dem Zentralnervensystem des Frosches bei Extraktion mit HCL-Alkohol und Trichloressigsäure sowie mittels Kochen. *Pflügers Arch. ges. Physiol.* 240, 449–457.

1938 (With H. HELLAUER.) Über das Acetylcholin in peripheren Nerven. *Pflügers Arch. ges. Physiol.* 240, 769–775.

1939 (With W. DULIERE.) Liberation of potassium by acetylcholine in the central nervous system. *Nature, Lond.* 144, 244.

1944 (With G.L. CANTONI.) Inhibition of cholinesterase activity of nervous tissues by eserine in vivo. *J. Pharmacol.* 81, 67–71.

1945 (With S. BLUMENFELD.) Digitalis and calcium. *J. Pharmacol.* 83, 96–99.

1945 The Edward Gamaliel Janeway Lectures. Aspects of the transmission of nervous impulse. Mediation in the peripheral and central nervous system. *J. Mt Sinai Hosp.* 12, 803–816; Theoretical and clinical implications. *J. Mt Sinai Hosp.* 12, 851–865.

1945 Chemical transmission of nerve impulses. *Science in Progress.* Ser. IV.

1946 The antagonism of tetraethylammonium against heart depressants. *J. Pharmacol.* 88, 136–141.

1948 Control of internal secretions. *Proc. R. Virchow med. Soc.* 7, 1–14.

1949 On the hypersensitivity of denervated structures. *Confinia neurol., Basel,* 9, 58–63.

1949 Antagonism between pressor and depressor agents in the frog's heart. *J. Pharmacol.* 96, 295–304.

1950 Smäsnak om farmakologi. *Med. Forum, Kbh.* 3, 225–234.

1950 Problems in the field of adrenal function. *Proc. R. Virchow med. Soc.* 9, 101–110.

1952 Ernst Peter Pick, the scientist. *J. Mt Sinai Hosp.* 19, xii–xiii.

1952 On the action of fluoride on the heart of Rana pipiens; preliminary note. *J. Mt Sinai Hosp.* 19, 1–3.

1953 (With E.P. PICK and J. WARKANY.) Alfred Froehlich, 1871–1953. *Science,* 118, 314.

1954 Introduction to Symposium on Neurohumoral Transmission, Physiological Society of Philadelphia, September 11 and 12, 1953. *Pharmacol. Rev.* 6, 3–6.

1954 Reflections on the study of physiology. *Annu. Rev. Physiol.* 16, 1–10.

1955 On the mechanism of the positive inotropic action of fluoride, oleate and calcium on the frog's heart. *J. Pharmacol.* 114, 90–99.

1955 Salute to Henry Hallett Dale. *Brit. med. J.* 1, 1356–1357.

1956 On the intraneural state of acetylcholine. *Experientia, Basel.* 12, 331–333.

1957 On the background of the discovery of neurochemical transmission. *J. Mt Sinai Hosp.* 24, 1014–1016.

1958 A scientist's tribute to art. Essays in honour of Hans Tietze. *Gaz. Beaux Arts,* 389–392.

OTTO LOEWI

F. BRÜCKE

In: Österreichische Nobelpreisträger für Medizin, Physiologie and Chemie
(Ed. O. Zekert) Heilmittelwerke – Jahrbuch 1961, pp. 65–85
Translation: Dr. Dennis F. Sharman (deceased March 10, 2002)

It is not an easy task to write an abridged study of Otto Loewi who received the Nobel Prize in Medicine and Physiology in 1936 for the brilliant proof that the effect of excitation of autonomic nerves on the effector is transmitted by specific neurogen substances, thus chemically. One of the difficulties in doing so is the fact that it is almost impossible to portray a life which contains academic achievements of fundamental importance, without deep respect which demands a certain restraint in the description. This, however, would deprive the portray of a true-to-life and versatile personality of vivaciousness and colour. If I try, nevertheless, apart from important biographical data to imagine the character of my adored teacher and fatherly friend, it is accompanied by flowers, by anecdotal tales which I heard partly out of his mouth, partly from his far too early deceased wife, but also by almost always cheerful episodes which I experienced myself.

O. Loewi, who would be rightfully indignant, if (like it happens every once in a while) his name was spelled wrongly, was born on June, 3rd 1873 in Frankfurt am Main as the son of a Jewish merchant. At the moment of his birth the affiliation to Judaism did not make a crucial difference within the liberal bourgeoisie of his home town. Especially in Frankfurt, similar to Prague, Jews who felt completely as Germans, took part widely in the intellectual and academic life. One just has to think of the excellent neuropathologists Weigert and Edinger and especially of the outstanding Paul Ehrlich.

Loewi attended the old humanist Goethe-Gymnasium and one young teacher was the – later on very famous – Philologist Karl Reinhardt. In his lifetime he considered the classic, humanist education as the only appropriate precondition for a career as a natural scientist. Moreover, it is possible that this formed his outspoken preference for teleologic thinking and his dislike of purely mechanical ideas in biology. He used E. Brücke's remark 'that teleology was a lady no natural scientist could live without, but who he did not like to be seen with in public.'

He did not feel very close to the religion of his fellow-Jews, possibly because of insufficient education, but his thinking was thoroughly theistic and far away from 'Freidenkertum' of some modern natural scientists. In his youth he became interested in any kind of cultural life: He pointed out important works of art history to each of his students. He adored especially for example the wonderful Velasquez-biography by Justi, but he also stayed in contact with modern art and preferred being surrounded not only by biologists of all schools but also and especially by artists. Music virtually excited him. In company he also loved to read out and discuss the best German lyric poetry and passages of books which impressed him.

The tendency to form things artistically was set off clearly in his essays and lectures: No matter on which occasion, he always tried hard to accomplish the best not only intellectually but also verbally.

He decided to study Medicine but since there was no University in Frankfurt he went to Strasbourg where he spent his student days, except for the third year when he studied in Munich. Like many German students at the end of the last century, he joined a 'schlagende Burschenschaft'. One can say that Loewi never fell for parochial nationalism and later on reached an attitude not dominated by the idea of nations; he constantly refused to take advantage of high academic rewards for national propaganda. Nevertheless, in spite of all his love for Austria, his attitude was one of a nationalistically feeling German professor until this was made impossible to him.

He did his dissertation with Oswald Schmiedeberg in Strasbourg. He told how naive he felt in this world-famous institute and how little he felt related to the topic which the unapproachable master gave to him. In 1897 he made lifelong friends in the laboratory with the ingenious and merry Walter Straub, later on professor of pharmacology in Munich. After the state examination he worked with Franz Hofmeister, a brilliant biochemist in Strasbourg for several months.

At this time, moods and events must have changed quickly, because Loewi did not stay long in the same place. Since his parents lived in Frankfurt, it seemed reasonable to him to work there with Paul Ehrlich at least for some time. After repeated meetings with Ehrlich, however, he had second thoughts about it because of the risk of possible mutual misunderstandings.

It, finally, seemed as if he wanted to leave the experimental medicine, when he entered the clinical for internal medicine of C. v. Noorden as an assistant. But his relationship to the clinical medicine was ambivalent. Suffering people caused so much sympathy, even anxiety in him that he, even later on, could not deal with them uninhibitedly. Furtheron in his life he could not have, for example, carried out an injection to somebody, even writing recipes, something he had to teach anyway, appeared to him as an almost unbearable responsibility and he never made out one for himself, but left it to his colleague.

Loewi was especially intimidated by mental diseases. An epidemic of severe pneumonia of which numerous patients died at his section and the painful feeling not being able to help, finally, made him leave clinical medicine completely. In this situation it must have been a relief when in 1898 H.H. Meyer asked him, to join him as an assistant in Marburg a. d. Lahn. He saw in Meyer a teacher whom he admired deeply till his hold age, even when his own academic achievements outgrew those of Meyer. The strict but cosmopolitan and sophisticated personality of Meyer, his immense knowledge, which enabled him – through his text book – to become 'praeceptor mundi', but especially the aesthetic enjoyment of working next to such a man, impressed all of his students deeply and gave them the opportunity to get acquainted with the most outstanding representatives of their field from all countries.

Apart from entering Meyer's laboratory, which after Meyer's appointment to Vienna in 1904, developed even bigger and more important with Loewi's help, an educational trip to England was of greatest significance to the young outside lecturer. Here he found the physiological line of pharmacology, as represented by Meyer, to be flourishing. The young generation of English scholars of this time impressed him deeply and among them he made lifelong friends with Ernest Starling and H.H. Dale. Dale told that Loewi was bubbling with ideas at this time and once as Dale corrected several wrong expressions, he said: 'Please, Dale, don't correct me, I do not wish to speak English correctly, I want to speak it quickly.' Inter alia, he got acquainted with Dixon and Elliott and among the books which influenced him most throughout his life was the General Physiology by Bayliss. His idols among the scholars were the German physiologist C. Ludwig, whose bust was standing in his room in Graz, and the Swiss Friedrich Miescher.

At this point of his life when he was highly open to everything new, but after all had found his own profession and his own way, he was impressed by Vienna. We as the younger ones will never ever be able to experience the magic of this endangered but still intellectually highlighted city. It was the city of the young Hofmannsthal, of Klimt and Loos, Arthur Schnitzler and Peter Altenberg, the city of Gustav Mahler, over which the glamour of the last years of a multiracial state was poured out and in which the best heads of half a continent gathered together – and it cannot be called back. The Danube Monarchy may appear to be cursed to revolutionary nationalists: Vienna, the Vienna of them all, was blessed.

Loewi could not forget his first evening in Vienna when he gathered with a number of young doctors and scientist in front of the Hotel Regina. Even 50 years later, as Honorary President of the Fourth International Congress of Biochemists, he was sitting in the same place and described that time, as if it had been real only yesterday. The city appeared to him like a fairy tale, the colleagues who were about his age educated, astute and quick-witted. During this evening he made lifelong friends with the brilliant heart-specialist Rudolph Kaufmann who later on founded the 'Verein Herzstation' together with Meyer. In Vienna he also met E.P. Pick again, whom he had already seen in Strasbourg.

At this time the outer development and the career of the young pharmacologist came along quickly and without any frictions. In the few years between his promotion (1896) in Strasbourg and his postdoctoral thesis (1900) in Marburg mainly studies about the metabolism were published, probably under the influence of Naunyn and Minkowski: First about purine and phlorhizin, then, however, mainly about the function of the kidney. Even if these studies, nowadays mainly interesting in a medical-historical sense, they still contain numerous thoughts and observations which are still relevant today. For example that the increase of urine by phlorhizin does not appear primarily, but as a consequence of the bond of water due to excreted sugar. Shortly after the postdoctoral thesis a study was published, on which Loewi still emphasises rightly today. It contains the proof that it is possible to keep dogs in a balance of nitrogen not only through the supply of protein but also through a mixture of amino-acids. Nobody had succeeded so far, because not all of the amino-acids were known and, moreover, the highly cleaned mixtures did not contain vitamins. Loewi's proof was stunningly simple: The egg of Columbus. He broke down animal organic protein by pancreatic enzymes so far until no peptide bond could be found in the Biuret reaction and fed the dogs with this material as the only N-carrier. They perfectly remained in a balance of nitrogen. This study is characteristic for Loewi's way of working. He loved quite simple, but convincing methods. He used to call them 'expeditive methods'. He was not very fond of technical appliances, although he could admire them in the hands of scientists who mastered them and used them to solve their problems. His clear reflections demanded an intensive intellectual work, which reminded of physical work, because he groaned often due to his efforts. In Vienna he became, according to old servants of the institute at the time of Meyer, a strict first assistant who demanded order and punctuality: now his attention was also drawn to topics about the effects of poison on the autonomic nervous system. This is an involuntarily functioning control-system whose effects are so very similar to the effects of several plant and animal poisons that they have to grip the attention of pharmacologists of now and then. The discovery of adrenaline and the solution of its structure in which also Viennese scientists, namely Otto von Fürth, took part and which Elliott masterly worked on in England around this time, encouraged the whole movement. Loewi was appointed associate professor in 1906 and in 1909 he already reported extensively about the pharmacology of the 'Wärmehaushalt' and finished his study on metabolism with a big summary in 'Noordens Handbuch der Stoffwechselerkrankungen' where he reviewed the complete literature on the effects of poison on the metabolism. Later on, he never composed bigger overall reviews again. A further example for his simple but well designed methodology may be this one by the end of his time in Marburg: Loewi found out that caffeine widens the blood vessels of the kidney which caused the venous outflow light-red-like. He wanted to know, whether this vasodilatation was the reason for caffeine diuresis. The problem was simply solved by putting the kidney in plaster which prevented its increase of volume. Even unter these conditions caffeine showed its effects and today we know that an increase of the glomerulus filtration rate as a consequence of an increase in blood flow, in fact, is not decisive for the formation of urine. A clinical observation dates from the time in Vienna which was named after Loewi: putting drops of an adrenaline solution into the eyes of diabetic patients causes dilation of the pupil.

In 1908 he married the daughter of the well-known organic chemist at the university of Vienna, Guido Goldschmiedt who originally came from Trieste. Those who knew this inwardly noble, highly educated woman, who even in hard times of persecution never lost her strength and who deeply loyal towards her famous husband endured everything that came along with the marriage to a scientist whose whole life revolved around his profession, would admire her almost as much as her husband.

Shortly after the wedding (1909) Loewi was appointed to Graz. Obviously, there was a certain resistance in the faculty there against the appointment Loewi's as a non-baptised Jew; but Meyer supported his excellent student and threatened to leave Vienna to follow a call from Berlin if Loewi would have been refused because of unobjective reasons. Later, at about the time when the state awarded to the quite young professor the title of a Hofrat, the integration was complete.

In a small, compared to modern standards, primitively equipped institute, Loewi began his work: the last publication in Vienna (together with A. Fröhlich) was the – even today unsolved – observation that cocaine intensifies and prolongs strongly the effect of adrenaline on smooth muscles. Now Loewi and his students (Gaisböck and v. Konschegg) worked on the effects of the vagus nerve on the heart of frogs and mammals. By the way, the first study in Strasbourg about the effects of hydrocyanic acid, phosphor and arsenic was already performed on isolated preparations of frog-hearts. But the series of publications of 1911 and 1912 is very interesting in a medical historical sense because it introduces further fundamental studies. Then, Loewi worked on a paradoxon: There are plant poisons, like pilocarpine and muscarine, which apparently imitate the effects of the vagus nerve themselves. But it was proved that small doses of these poisons made the heart insensitive to electrical stimulation of the vagus. Now Loewi showed that also the other way round an electrical stimulation of the vagus nerve weakens the effectiveness of these substances on the heart. Although the conclusion which seems obvious today was not drawn from it that exactly because of this stimulation of the vagus a peripherally effective substance is released which joins in a competition with plant poison for a receptor molecule on the muscle, Loewi came very close to the solution of this problems which he published 10 years later. Anyway, he noticed that pilocarpine and muscarine had to have a purely peripheral effect at the transition of the nerve to the muscle. Another important diagnosis which was later on resolved as well, is the fact that physostigmin significantly increases the effect of an electrical stimulation of the vagus, but not the effect of pilocarpine and muscarine.

Another field Loewi worked on thoroughly later was the effect of inorganic cations on the cardiac muscle. Especially the results he achieved after the First World War on the increase of sensibility of the cardiac muscle for calcium under the influence of digitalis are still very important today. Today the most sophisticated methods of modern electrophysiology are used to work on the unsolved questions of this field still referring rightly to the pioneering work of Loewi in this field.

He himself often stressed that he did not judge a scientist by what he had already done but by what he would expect for him in the future. Looking back, one can see that a lot could be expected from the author of these studies on the heart. The problem he dealt with was always fundamental, the methods were simple, but the discussion of the results shows the most accurate knowledge of previously published studies and an astute sense to realise coherence.

A few years after these studies the First World War destroyed the results of a long period of peace, the old order in the state and in the society and most of the private fortune. Work and the world-wide connection of science turn out to be stable. During the war Herr Hofrat helped to change bandages in the surgical clinic. But soon after the end of the war, during the weird, turbulent twenties the whole world experienced an astonishing revival of intellectual productivity. In this time Loewi who had just turned 50, developed from an expectionally talented professor to a brilliant, pioneering scientist.

By this time he had settled down completely in Graz; he loved this city and its surroundings, he also loved its people. Even in the thirties, by the time I got to know him, he could suddenly interrupt his work in the middle of the morning and walking the Schloßberg up to admire the blooming Japanese cherry trees. He then told about the many clever and talented people who lived in his town and found appreciative worlds for all those different characters. He only hated middle-class mediocrity; in dealing with those people he almost played dead by fixing his light grey, slightly protruding eyes on them and by making completely absent minded remarks. After such a meeting I heard a simple minded chemist grumbling: 'One cannot talk with Hofrat Loewi, he doesn't even know what a one percent solution is.' His house, where Frau Hofrat regularly introduced 'jours', frequented not only for colleagues but also for assistants, artists and merchants, among them also several later National Socialists who then incomprehensibly blind declared their host to be a suspect and an enemy to the German people. A professor who was appointed to Graz told me, of course much later that he was warned of the dangerous Freemason. He, however, did not care about this stupidity, visited Loewi several times and was very astonished to meet almost all those who warned him of Loewi there as well. One of those guests allegedly said later: 'Loewi should have known how much his invitations would harm me!'

But in the twenties everything was still quiet in this concern. Loewi was very popular with his students as a stimulating lecturer and a mild examiner and a lot of young research workers from departments and clinics asked him for an academic advice. A number of new colleagues joined the Institute like A. Jarisch and later on H. Häusler, Loewi's current successor. A number of student-anecdotes stem from this time. One candidate, for example, answering the question where the iodine came from said: 'From the Iodine-tree!' Loewi replied: 'Then come again when the Iodine-tree is blooming.' Another time, Herr Hofrat came excited out of his room into the laboratory and said to an assistant: 'What shall I do? Look at me, here grows a goitre!' The assistant answered: 'Herr Hofrat, the best thing you can do is buy yourself a "Steirergewand", (the national costume of farmers of the region), because then the goitre won't be noticeable.' A few days later this story was printed in the Simplizissimus (a journal devoted to literature) and Loewi was indignant about it because he assumed an indiscretion. But when he found out that a poor student who was present by chance had received 50 Mark for this story, he forgave him immediately.

Working in this peaceful atmosphere, Loewi made the first step in a series of revolutionary discoveries in 1920. In 1903 already, 17 years earlier, he discussed with a co-worker the possibility that the excitation of autonomic nerves is transmitted chemically into the corresponding organs and is responsible for their reaction of the stimulation of nerves. Since he did not know a way to prove this chemical hypothesis, he deliberately did not deal with this problem anymore. He lost this indifference from one day to the other in 1920. One night he dreamt about an ideal methodical solution of the problem, woke up and jotted down some notes. The next day he forgot both the dream and the associated thoughts, but he could not decipher the hasty notes he made either. In the following night the dream reoccurred and he was completely aware of it. He got up immediately and had already performed the crucial experiment all by himself in the institute when the assistants turned up at 8 o'clock. In March 1921 he published four pages in Pflügers Archiv as the first of his eleven classic notes on the humoral transmission of the effects of cardiac nerves.

Before we go more into detail, as far it is possible in such a short sketch, it must be stated that, right from the beginning, Loewi was aware of the special significance of his work. This becomes evident not only in personal notes but also in the way it was published: Loewi published all his studies alone except one which was published by Witanowski alone and the last one which he published together with Navratil.

He did not choose the pharmacological archive, where his earlier studies and those of his students had been published, to do so, but Pflügers Archiv, a far spread and highly respected physiological journal. The studies are shortly written and no word is unnecessary: the literature which Loewi normally takes into account is greatly omitted, because there are – to be exact – no real precursors. It is right that later on several alterations were found made by English authors, especially Elliott and Dixon which proves that they thoroughly dealt with the idea that the similarity of effects of autonomic nerves with those of plant poison may be based on the production of such substances in the body of animals. Especially Sir Henry Dale referred to such spots and to the fact that Du Bois Reymond had already thought of a chemical transmission of impulses on striated muscles. But none of these authors could provide an experimental proof for such hypotheses and in 1920 the view was prevailing that the excitation of the effector of an autonomic nerve was caused by the action potential which appears during the excitation, thus being electrically. Only today, we know that these potentials would be far too weak to cause such effects. The authors mentioned were even less aware of the chemical structure of those possibly effective substances.

It is also remarkable that several years later Loewi replied to an article which was completely erroneous and unjustified critical, in a polemically sharp way which was quite unusual for him. In fall 1921 already he shortly presented his discovery at the Congress of the German Pharmacological Society and extensively at the Congress of the 'Deutsche Naturforscher and Ärzte' in 1924. On this occasion he repeated the main results so often until the chairman v. Frey, finally, said: 'Mister Loewi, we have already understood you!' Loewi excused himself making the remark that his astonishing discovery seems so remarkable to himself and he made himself aware of it again and again.

Even after Loewi's first publication, it is certain that it took most of the physiologists and pharmacologists several years to grasp what it signified. Admittedly, someone uninitiated, who has not been present during the experiment, could not necessarily be convinced by the first records published. Only five years later, for example, the Prague physiologist Kahn repeated very aesthetically Loewi's basic experiment with a useful, purely technical modification and from now on nobody could doubt it anymore. Due to these circumstances, Loewi had the opportunity to examine everything essential by himself.

What was the basic experiment then? Loewi prepared two frog hearts which he isolated from the body and which beat against a cannula filled with a physiological salt solution. Now the nerve was stimulated electrically for a while until the heart beat became slower and slower and finally stopped. The content of such a period of stimulation was transferred into the second heart (receiving heart), which had been emptied before and it showed, although weaker, all the details of the inhibition as it had happened with the first one. During the period of stimulation a substance with a vagus-effect (vagus-substance) was released to the physiological salt solution and this substance could be transmitted humorally to the receiving heart. The formation of such a cardiac inhibitory substance explained for the first time the paradoxon Weber has already observed in the middle of the previous century: more stimulation of the nerves causes less work of the heart: a chemical substance was formed! Right from the beginning it was clear to Loewi that each attempt to chemically define the structure of the vagus substance had to fail because of minimally available quantities and he did not spend time on that. He rather followed ist reaction to simple operations: heating up with additional acid and 'Alkali' and so forth. His precise knowledge of known substances with similar effect was especially helpful to him. He assessed that it could only be the ester acetylcholine. One gram of this substance would have to be diluted with so much water as a train with 40 cars could carry in order to reach a concentration which is clearly effective on the frog heart. Even though in the meantime a big number of new esters of the cholins had been synthesised and in the early twenties acetylcholine had not been proved to be existing in the animal organism, all experiments conducted since then supported Loewi's assumption, because on several examined testing objects the vagus substance constantly had the same quantitative effect as acetylcholine. There Loewi made another fundamental discovery: He found out that the vagus substance and acetylcholine are deactivated the same way by an extract of the cardiac muscle, an enzyme, the acetylcholinesterase. The biologic significance of this newly discovered enzyme, as Loewi proved, lies in the fact that, obviously, acetylcholine which is released by a local excitation of the vagus cannot cause a long-distance effect. Otherwise, we would suffer simultaneously from a slowing down of the heart and for example also diarrhoea or contraction of the pupil. Loewi also proved that the vagus substance, in fact, may cause all details of the heart effect of an electric stimulation of the N. vagus which makes it alone responsible for all those effects. Later on, this was proved especially by Rothberger for the mammalian heart. Another of Loewi's fundamental findings (together with Navratil, 1924) was the idea that the alkaloid physostigmine inhibits the acetylcholinesterase and consequently extends the effect of the vagus substance and the excitation of the vagus, whereas this is not possible, as he himself found out already in 1911, with the alkaloids pilocarpine and muscarine which are not cleared by cholinesterase. For the first time in the history of pharmacology, this result explained the mechanism which makes an alkaloid effective. It has to be mentioned that Loewi determined as well that the paralysing effect of atropine on the vagus did not result from a smaller production of the vagus substance, but from the fact that the cardiac muscle became insensitive towards this substance. From the first moment of this discovery on, he did not doubt that the discovered causative substance can be found as well on all other endings of parasympathetic nerves and this not only after electric stimulation but also in the course of physiological series of stimulation. A fundamental contribution was made by Engelhart from Loewi's laboratory (1931): The destruction of acetylcholine in rabbits was prevented by appropriate doses of physostigmine. Now one eye was kept in darkness and the other one exposed to intensive light. On the exposed side a strong contraction of the pupil happened due to the light-effect by parasympathetic nerve fibre. Examining the anterior chamber fluid of this side, it contained much more acetylcholine than the one of the darkened side. Here, the causative substance was released under completely physiological conditions.

Here we now stop describing Loewi's fundamental discovery and just stress one more time that the method of using an isolated frog-heart was not only very useful to detect the created substance, but also, what is particularly important, to determine ist quantity. By the way, Loewi found that the effect of the sympathicus (the effect of the nerve which accelerates the heart) is humorally transmitted as well, and in a study he conducted much later, he defined adrenaline as the transmitting substance for the frog-heart. At the begin of the thirties several experiments showed the release of acetylcholine at various effectors of the parasympathicus, which have been completed ever since. H.H. Dale and his colleagues discovered that acetylcholine is playing the role of a transmitting substance also by stimulation of the motor nerves, innervating the striated muscle and the preganglionic fibres of the autonomic nerve-system. This indicated a wide extension of the humoral (chemical) transmission discovered by Loewi. Especially in such sites as the motor endplate or the synapses, the local concentration of the acetylcholinesterase played an essential role, because the explosively released amount of this substance has to disappear again in a fraction of a second, otherwise it would be paralysing. It took a long time until the discoveries and their meaning became widely accepted and some details are still controversial today. It is obvious, however, that no modern discussion about the idea of excitation at the excitable membranes of nerves and muscles can ignore the theory of chemical transmission of excitation. There's still a lot to do in the future especially in explaining most complicated excitation of the spinal cord and the brain. But even there acetylcholine is playing a role as far as today's knowledge goes. It is a fact that throughout the last 30 years all discoveries Loewi's were confirmed and that they (like all brilliant ideas) will be of importance to the future research.

Shortly before finishing up the most fundamental experiments on the 'vagus-substance' a scientific mishap happened to Loewi, which I only mention due to his reaction on it. Throughout several years he conducted experiments together with a colleague on a simple explanation of the mode of action of insulin; which especially interested him because he was particularly interested in problems with diabetes which he had followed since his time as assistant. By 1928 some studies were published first in Sweden which could not reproduce the results which have been made in Graz. Finally and for unexplainable reasons, even Loewi and his colleague were not able to achieve the same results as in past years although some independent researchers had confirmed them before. Instead of just letting 'fall asleep' those series of experiments, as it usually happens, Loewi decided (1929) in an exemplary publication, and because of that worth, to revoke his earlier results himself and to provide a complete description of all the facts, which definitively was awkward to him. This all happened in a moment in which confessing a mistake could make difficult for a new big discovery to be acknowledged.

At the beginning of the thirties, Loewi could make sure with Engelhart, that the destruction of the acetylcholine in the blood happened by an esterase, which could be blocked by physostigmine. Only very much later they found out that it was another enzyme than acetylcholinesterase. Again, it's remarkable, that right away Loewi was far more interested in the tissue esterase, which seemed physiologically much more important. In 1931 the study of Engelhart about the formation of acetylcholine in the eye was published, and then the big field of chemical transmission of the effects of nerves has been finished as far as Loewi's own work is concerned. It was a satisfaction to him, to see the fundamental significance of his studies emphasised on the occasion of the Nothnagel-Lecture by his friend Sir Henry Dale in 1933, who had already started to prove the various chemical transmissions at all kinds of nerve-junctions and later on, in 1953, summarised this whole study in a lecture. Loewi, however, seemed to feel that he himself would not exceed his great discovery. This did not leave him passive at all, but he restricted himself to support and stimulate young colleagues. Another fundamental discovery was the fact that the much available liver-glycogen of the hypophysectomiced frog could not be transformed into glucose. Furthermore, Loewi was the first one who used the fluorescence-method discovered by Paget to measure even smallest quantities of adrenaline in frog hearts while electrically stimulating sympathetic nerves.

He knew that friends proposed him to the committee of the Nobel Prize and it was clear to him that his discovery as a unique and coherent study was especially appropriate for the Nobel Prize in Medicine and Physiology because it supports progress like hardly any other which became more obvious following Dale's studies and those of his colleagues. Each fall he eagerly waited for the decision made in Stockholm, until, finally, in 1936, the Prize was awarded to him and Sir Henry Dale. One can hardly imagine the family's excitement, joy and pride: One has to remember that 1934 as a very fateful year for Austria has already passed. Political fanatism was creating quite a stir already. Even more so, this highest academic award gave Loewi great satisfaction which made a big impression everywhere in Austria. By the time Loewi gave his wonderful speech in

Stockholm, nobody could imagine that Germany's regime in power would forbid its citizens to accept this prize which was awarded by objective, academic points of view only. What made the committee even more honourable was the fact that this did not keep them from still awarding this prize to Germany's citizens (G. Domargk and O. Hahn). In previous years, Loewi has already been invited to very honourable lectures. In 1933 to the Dunham Lecture in Harward and to the Harvey Lecture in New York. In 1935 he gave a lecture at the Royal Society in London. In 1936 he was made Honorary Member of the Royal Society of Scotland and later on (1944) received the Cameron Prize of the University there. In the same year he was elected as Honorary Member of the New York Academy of Medicine and in Austria he was awarded the 'Grosses Ehrenzeichen für Kunst und Wissenschaft'.

Due to his deep optimism, his inexperience in political affairs and his convictions of always having been useful for his country, Loewi peacefully and actively spent his last two years in Austria. The simple people, craftsmen, merchants but also the majority of students admired the famous scholar who enjoyed talking to them about his problems and was willing to listen and give advices. Each year the venerable Hans Horst from Vienna visited his home. He had no difficulties at the University until the catastrophe happened.

The following events occurring to Loewi could be almost passed over because they were so disgraceful, dangerous and decisive in Loewi's life. It seems as if he himself did not forget them but excused them and re-established his friendship with all who did not take part actively in the evil.

He was imprisoned but released after a relatively short time (two months). When I saw him the last time in Europe, he was sitting in the garden of his villa smoking several cigarettes. He lost a lot of weight, but he was in a good mood and joked how this arrest spared him from travelling to Marienbad to lose weight. Before he could emigrate he had to give up the Nobel Prize in favour of the Third Reich in the presence of Swedish representative, although Austria had allowed him to leave it in Sweden. He had to leave his wife all by herself in Graz. Everything was taken away from her, she was deported and had to stay in Viennese Hotels in which Jews were packed in afraid of being taken to death camps. I shamefully remember how scared I was to visit this woman who was totally calm und just happy that her family's flight had succeeded. Of course there were people who were not Jewish but fulfilled their duties as humans without fear.

Loewi was welcomed with open arms in England and then in Belgium, where Professor Henri Fredericq, one of the last Grandseigneurs of European science, provided him with the Franqui Professorship in Brussels from 1938 to 1939. At the beginning of the war, Loewi stayed with friends in England. There he received an invitation to come to the US and work as a research professor at the New York Medical School. He happily accepted this invitation, especially, in order reunited with his wife again.

He told that he had some unusual difficulties at the American consulate as they asked for an official document he could not provide; they even denied accepting Sir Henry Dale as a witness. Finally, Loewi said: 'But, Mister Consul, you can look up the "Who is Who"!' Now the desperate officer cheered up: 'Oh yes, if you are in there, it definitively will be a difference!' And he provided the desired visa in a very friendly way. Loewi was about to leave but turned around and said: 'And, Mister Consul, do you know who wrote the article in 'Who is Who'? I did it myself. Good by, Mister Consul.'

Loewi arrived in New York in 1940 and lead a very moderate life even though his friends helped immediately. They started his professorship at the New York University Medical School which he still holds today. He can be proud of being the only American professor who is not subject to an age limit. He did not have to teach classes regularly but was invited again and again to single lectures under convenient conditions. For example as a Dohme Lecturer in 1940, to the John Hopkins University in Baltimore in 1941, as a Walker Ames Visiting Professor in Seattle in 1942, as a Janeway Lecturer in Mount Sinai hospital New York, as a Hughlings-Jackson lecturer to the McGill University in Montreal, Canada in 1944 and as a visiting professor. He published several studies about the effect of ions, but also his outstanding, thought out lectures: In 1953 the book 'From the workshop of discoveries' with the University of Kansas Press, Lawrence, in which he dealt with the question of medical education but also with the problem of 'discovery' in medicine.

He spent the summers mainly in a perfectly equipped and climatically most convenient Marine Biological station in Woodshole. There he could conduct experiments or talk with a few close friends, but most important he had access to the complete physiological, pharmacological and biochemical world-literature.

Loewi's extraordinary position in the American world of science was not solely consolidated by his past as a university professor or the Nobel Prize. The American people appreciated and loved his complete modesty, has spirit of taking lively interest in modern problems together with his outstanding skills in explaining, his optimism and his capacity for enthusiasm, not to forget his helpfulness toward everybody who asked him for his scientific advice. In addition to that, he knew numerous academic publications of this century in different languages and – due to his own experience and his discussions with the most influential experts – was able to point out the most important facts of progress or in problems which were presented to him. His knowledge and his remarkable memory had already been astonishing in Europe and did not decrease as he got older. Instead, they enabled him to follow new developments. His definite preference for social gatherings made him meet countless, mostly younger natural scientists in his small apartment in New York. He was one of the few European emigrants who not only had contact to German-speaking colleagues but also and right away with Americans of different backgrounds. Soon after the war some Austrians and Germans came looking for him and he did not have any embittered feelings towards them. The experience of a surgeon who came to see him is characteristic for that: Ringing the door bell, Loewi himself came to open the door, looked at him quickly and said: 'You must have been a nazi as well!' Answering truly the visitor said: 'Yes, I was a party member!' Now Loewi asked his wife to come to the door because, finally, a nazi visited them while all other similar guests denied having had to do anything with the party. Then he very friendly welcomed his guest and talked with him for a long time.

Due to his naturalness and openness Loewi was the very right person to create an new but unpaid profession: one of an academic mentor and partner in discussions, a so-called 'consultant scientist'. Everybody who works actively in the field of sciences knows how supportive those conversations can be for the own thinking. Basically, it is an old 'obstetric activity', as Platon has described already, an activity which is completely different from the actual teaching and learning but close to the English 'tutorial system'. Of course, only a few can succeed in this profession.

In June 1960, Loewi celebrated his 87th birthday. Accidents, surgeries and illness weakened him and the wrinkles in his face mirror the drastic experiences of his life. His spirit triumphed over his body. His lively letters and his conversations dominated by the presence prove it. Nothing justifies talking about this successful life by looking from the point of the past. An evidence for that is his journey to Vienna at the age of 85 in order to attend the Fourth International Congress of Biochemists and to give a wonderful speech there as the President of the Honorary Committee. This speech ended with some verses from a poem, written by the outstanding biochemist and Nobel Prize winner Otto Meyerhofer shortly before his death in 1951. Otte Loewi still identifies himself with the idea of this poem:

How wonderful, and far about all earthly feeling
There came to us the soul, and the soul's ripening.
Thus, slowly, from the man's true beginning,
We learned of God, and that He us created
To fulfill the holy work of Meaning.
For this He gave us our only calling:
To, filled with amazement, make the revelation
Of His secret, our intended gleaning.

Gar wunderbar, hoch über Erdensinnen
Kam uns die Seele und der Seele Reifen,
Daß langsam, von des Menschen Anbeginn,
Wir lernten Gott und daß er uns erschuf,
Das heilige Werk der Deutung zu erfüllen,
Daß er uns gab, als einzigen Beruf,
Voll Staunen sein Geheimnis zu enthüllen.

LOEWI'S TIME IN GRAZ

Appointment to Graz

The appointment of a professor at that time had to be approved by the Emperor. The predecessor of Otto Loewi left Graz in 1908 to become Professor of Pharmacognosy in Vienna. The Medical Faculty in Graz had to prepare a list of candidates and suggested that the chairs of pharmacology and pharmacognosy existing to date should be separated; the Ministry of Culture and Education did not agree to this for financial reasons. The chairs were, however, separated a few years later. Otto Loewi, at that time coworker of H.H. Meyer in Vienna, and Eugen Rost, Berlin, headed the list. In the report of the Minister to the Emperor, the personal view of the local representatives of the Crown on the character of the suggested candidates was given – a common procedure. Loewi's Jewish confession was addressed, but not regarded as a negative influence in view of his otherwise first class personality and reputation in research. The report of the faculty of March 2, 1909, which influenced the report of the Minister to the Emperor on October 30, 1909, was completed by the Ministry on November 2, 1909, and announced immediately to the faculty. Loewi took his oath on the constitution on December 2, 1909 and delivered his Inaugural Lecture on January 25, 1910.

The Department

In Graz, Loewi found the Department on the first floor of the impressive Medical Institute Building, erected around 1900, at the most prosperous time of the Monarchy: a staircase as in an Italian palazzo, large rooms laid out with oak parquet floors, steam heated, the professor's office with walls painted in Pompeian red and equipped with oak furniture. There were positions for three graduates and one 'servant' for everything including the catching of frogs and collecting the monthly salary for the professor and his staff from the accounting office. There was no secretary. The professor had to buy his own typewriter. A special room was set aside for an analytical balance. Otherwise, there was just enough equipment to set up an isolated frog heart and to record its movements on a clock-driven kymograph. The following years of war, the postwar economic problems and the years of the Great Depression did not bring any improvement. A desperate application by Loewi for an extra allowance of 425 Austrian schillings remained without answer. Tausk [1964] described the facilities in Graz as 'very restricted'. A motor-driven kymograph was a donation from the Rockefeller Foundation.

Coworkers (the Years in the Department Given in Parentheses)

Hans F. Häusler (1924–1938), who never joined the work on humoral transmission, worked on an insulin-antagonistic factor. In the following years, he held a position in the pharmaceutical industry, and returned to Graz as Professor of Pharmacology from 1945 to 1967.

Theodor Konschegg (1911–1913) was a coworker of Loewi in the experiments on the influence of calcium ions on the effect of digitalis [Loewi, 1918]. Thereafter, he changed to pathology and held the Chair of Pathology in Graz from 1945 to 1956.

Adolf Jarisch (1918–1924) became afterwards Professor of Pharmacology in Innsbruck from 1924 to 1958.

Carl F. Cori came from Prague and worked as tutorial assistant for one year (1921–1922), whereupon he left for the USA. He and his wife Gerty Therese were awarded the Nobel Prize in 1947 for their work on the glycogen metabolism.

Ernst Navratil (1924–1927) collaborated with Loewi on the actions of atropine and physostigmine, changed thereafter to obstetric surgery and was head of that clinic in Graz (1947–1973).

Erich Engelhart (1928–1931) made the important discovery that acetylcholine is not only released by electrical nerve stimulation, as common in the experiments performed, but also by a physiological reflex reaction as response of the pupillary constrictor to stimulation by light in the eye; it was the only publication in this field without Loewi as a co-author, and therefore often missing in the list of publications [Engelhart, 1931].

Franz Brücke (1932–1934) later became Professor of Pharmacology in Vienna from 1945 to 1970.

Horst Hellauer was the youngest and last coworker of Loewi in Graz, working on the acetylcholine content in peripheral nerves. These experiments found a dramatic end. Loewi, already in jail in March 1938, submitted his recent results on a postcard for publication. This was mentioned by P. Rosbaud [see Cahn, 1994], later the Scientific Director of Pergamon Press: 'In 1938 I was Research Consultant of the Springer Verlag in Berlin. In the mail room I found a postcard to the Editor of *Naturwissenschaften* written in pencil:

"As I don't know whether I can continue my research work in the future, I would like to have my recent results be published briefly. Prof. Dr. Otto Loewi, Graz, Paulustor-Gefängnis."

With agreement of Dr. Springer, I immediately went to London and presented this distressing (unfortunately lost) document to Sir Henry Dale who presented it at the next Meeting of the Physiological Society.'

The chapter would be incomplete without mentioning two contemporaries at that time, Wilhelm Berger and Marius Tausk.

Wilhelm Berger – the First Dissident

From 1931 on, Professor W. Berger was Head of the Clinic of Internal Medicine in Graz. His scientific interest was mainly in the field of allergy and he edited the *Textbook of Allergology* (jointly with K. Hansen). Besides his scientific reputation he was known as a very careful clinician. His character was described as conservative, but he was keeping his distance from contemporary political affairs. There was only occasional contact between him and the Loewi family. This changed when Otto Loewi had a severe renal colic during his imprisonment and was transferred to Berger's Clinic. Mrs. Loewi was informed about the patient at Berger's private surgery, whereupon she had tea with Mrs. Berger; this was done in a confidential atmosphere because any contact with Jews was already forbidden.

Berger was one of the very few to foresee the coming events. In the summer of 1939 he left Graz with his family on a trip to Italy, and went from there to Switzerland and the USA. The start there was not easy because he had to repeat all medical examinations according to the rules of the American Medical Association, which he, who had received his Doctor Degree at Innsbruck sub auspiciis imperatoris, merely regarded as a refresher course. No wonder, because the sub auspiciis imperatoris award was given only to those few students who received the highest marks in all examinations at the gymnasium and later on at university. Soon thereafter he was professor at the New York University Medical School and gained a high reputation as a consultant in gastroenterology.

When Loewi and his family arrived in New York in 1941 a close and continuous friendship developed with the Berger family [Braunsteiner, 1969].

Ulrich Weiss, Otto Loewi's son-in-law, wrote: 'The Bergers left their home country in firm confidence, but without being forced to do so, like the Jewish emigrants. How desperately difficult such a decision was, even when no other decision was possible – this, I know from my own experience.'

Marius Tausk – from Student to Research Director

Marius Tausk was born on February 12, 1902, in Sarajevo where his father, Dr. jur. and Dr. med. Viktor Tausk, was Government Official of the Austro-Hungarian Monarchy. A few years later the family moved to Vienna. Marius studied medicine in Graz from 1920 to 1926. Shortly after his graduation in January 1926, his mother, who was the secretary of the well-known Socialist Victor Adler, arranged his participation, together with a group of young socialists, at an International Meeting in Amsterdam. When Tausk mentioned this incidentally to Loewi, he gave him a letter of recommendation to his old friend Ernst Laqueur, Professor of Pharmacology in Amsterdam. At the resulting visit Laqueur invited Tausk to join his group as a guest worker. He wrote in his memoirs: 'What a contrast between the small institute where I had attended Loewi's classes in pharmacology and Laqueur's palace were I landed in 1926!' A few months later Laqueur found a job for him in a small pharmaceutical company in Oss, called Organon, which was engaged in the production of hormones. Tausk remembered: 'I started my work with no training or experience whatsoever to qualify me as the head of even a minor production unit.' He joined the company in November 1926 and remained with it for the rest of his life. Under his guidance, Organon developed into one of the world-leading companies in the production of hormones. In 1926, Otto Loewi could not have foreseen that 12 years later, after his release from prison, Marius Tausk would pave his way to Sir Henry Dale in London. Tausk was awarded the honorary doctor's degree of the Medical Faculty of the University of Graz in 1972.

Successors

After Loewi had been forced to leave the department in spring 1938, it was difficult to find a successor. Professor F. Hildebrandt, Giessen, found 'the facilities in Graz so disappointing that he would need years to raise it to the standard of his Department in Giessen'. Kuschinsky, after having returned from a long visiting professorship in China, was appointed to Graz by the Reichsminister in Berlin, but left after 3 months for the excellently equipped department in Prague. Years of improvisation followed until Hans F. Häusler arrived in 1945 and held the chair up to his retirement in 1967.

My own entrance into pharmacology was not planned and was the result of a few lucky incidences. As a student I served, from 1946 to 1947, as tutor in the severely bombed Physiology Department. After graduation I was offered a job in pharmacology, which was not much better. Only the professor's office and one other room, used for everything else, including the keeping of rats, were heated. We spent our time with primitive tasks like producing copper ion-free distilled water and crystallizing sodium chloride to reach a purification not killing a frog's heart. I desperately looked for a job appropriate for a young doctor. At that time an Official of the World Health Organization turned up in Graz and was immediately brought to Professor Häusler, the only English-speaking professor of the faculty. I am very grateful to him as he persuaded the WHO Official to offer me, although still without any experimental experience, a fellowship of 6 months and arranged a place for me in Sir John Gaddum's department in Edinburgh. My arrival there, 1 year after my graduation in 1947, was my entrance into pharmacology and decisive for the rest of my life. After having held the Chair of Pharmacology in Tübingen from 1961 to 1969, I held the chair in Graz from 1969 to 1992 (F.L.).

References

Braunsteiner H: In memoriam Prof. Dr. Wilhelm Berger. Wien Klin Wochenschr 1969;81:529–530.

Cahn RW: The origins of Pergamon Press: Rosbaud and Maxwell. Eur Rev 1994;1:37–42.

Engelhart E: Der humorale Wirkungsmechanismus der Oculomotoriusreizung. Pflügers Arch Physiol 1931;227:220–234.

Loewi O: Über den Zusammenhang zwischen Digitalis und Kalziumwirkung. Arch Exp Pathol Pharmacol 1918;82: 71–158.

Loewi O, Hellauer H: J Physiol 1938;93:34; cited in Lembeck F, Giere W: Otto Loewi: Ein Lebensbild in Dokumenten. Berlin, Springer, 1968, p 57.

Tausk M: Organon – the Story of an Unusual Pharmaceutical Enterprise. Oss, Akzo Pharma bv, 1964.

COMMENT ON LOEWI'S DREAM

In my mind as a pharmacologist, a dream is a cerebral function in which stored facts, speculations or emotions reach the level of consciousness inducing the arousal of fantasy or the motivation for an action. Concerning the discovery by Otto Loewi, my 'dream analysis' points to the following facts stored in his mind:

(1) He had already earlier used the method of the isolated frog heart, on which his experiments were based. Fröhlich and Loewi used it to show the 'potentiation' of the effects of adrenaline by cocaine. It helped Loewi and Konschegg to analyze the interaction of calcium and strophanthin.

(2) Otto Loewi, although professor, was able to perform experiments on the frog heart with his own hands, which is not very common today.

(3) No license was needed to collect frogs in the vicinity of Graz. In order to save this 'protected species', a special permit of the Ministry of Science and Education would be required today.

(4) The professor was not engaged in many committees dealing with administrative, political or social problems. He could, therefore, attribute much more space in his brain to his ideas about research and his imagination devoted to new experiments.

(5) The professor was hardly disturbed by many phone calls and E-mails.

(6) Otto Loewi had earlier findings readily available in his mind, such as the parasympathomimetic actions of muscarine and eserine, the muscarine- and nicotine-like effects of acetylcholine [Dale, 1914], the emergency function of the adrenals, the parasympatholytic action of atropine, that is all the earlier approaches to explaining the neurotransmission [Du-Bois-Reymond, 1877; Elliot, 1905]. Loewi also speculated about the possibility of chemical neurotransmission years before performing the crucial experiments. Koelle wrote in his careful description of the earlier findings in this field: 'Following his return to Marburg in 1902, during a walk through the woods with Walter Fletcher, he expressed the thought that the vagus nerve might release a muscarine-like substance to produce its cardioinhibitory effect' (see p. 104).

The essential key of Loewi's design of his experiment was the *transfer* of Ringer solution collected from one heart under electrical nervous stimulation to another, the 'test heart'. In those days the isolated frog heart was the only preparation available to demonstrate effects of very small doses of acetylcholine and adrenaline. This transfer was possible due to the invention of a suitable cannula by Straub and the previous introduction of the isolated organ bath by Magnus.

Facts and findings about memory formed the intellectual background which we might regard as a 'dream'; this could be so intense as to initiate an immediate action. Christiane Vulpius expressed this notion in a simple way in a letter to Goethe: 'I had dreams about you all the night, because I spent hours in thoughts about you all the day long' [Damm, 1998].

Otto Loewi's dream has frequently been quoted. A comprehensive description was published by Ulrich Weiss, his son-in-law, and Ronald A. Brown [1987] (see p. 74).

References

Dale HH: The action of certain esters and ethers of choline, and their relation to muscarine. J Pharmacol Exp Ther 1914;6:147–190.
Damm S: Christiane und Goethe. Frankfurt am Main, Insel Verlag, 1998.
DuBois-Reymond E: Curare. La science expérimentale. Paris, Baillière, 1878, pp 237–315.
Elliot TR: The action of adrenaline. J Physiol 1905;32:401–467.
Weiss U, Brown RA: An overlooked parallel of Kekulé's dream. The discovery of the chemical transmission of nerve impulses by Otto Loewi. J Chem Educ 1987;64: 770–771.

AN OVERLOOKED PARALLEL TO KEKULÉ'S DREAM: THE DISCOVERY OF THE CHEMICAL TRANSMISSION OF NERVE IMPULSES BY OTTO LOEWI

ULRICH WEISS RONALD A. BROWN

Reprinted with permission from J. Chem. Educat. 1987;64:770–771

Much controversy has centerd recently on the reliability of Kekulé's late (1890) account of the dreams that led him to recognise the ability of carbon atoms to form chains (1855), and to formulate the structure of benzene (1865/66). This debate [1] has generated a great deal of interest. The verdicts [2] range from: no reason to disbelieve (what should the world-famous elderly man have expected to gain from concocting such a fairly implausible story?) to the claim that the illustrious chemist, 'in the twilight of his life', had acted from German chauvinism [3a], and the recent surprising assertion that Kekulé's account was given 'at a beer party, when he likely had had one too many' [4].

Three distinct topics are at issue: the trustworthiness of Kekulé's account, the question of whether scientific ideas can appear in a dreamlike state, and the wider one of the role of intuition and imagination in science. Much of the negative view on the first one of these questions is based on the fact that Kekulé had not mentioned these dreams *in print* (emphasis ours) between their alleged occurrence and 1890. Coupled with this is the denial that nondeductive intuition could play any role in the advancement of science, which is said to be possible exclusively through experimental work to get hard facts first, and the stern injunction that chemists must stop dreaming (i.e., having any ideas?) and must 'get out from under the Kekulé myth' if 'they wish to succeed' [3].

In view of all these conflicting interpretations, it is surprising that hardly anyone [5], to the best of our knowledge, has mentioned a closely analogous instance of a dream with important scientific consequences, an instance that has many similarities with Kekulé's dreams but that, in contrast, is as completely documented as such an entirely subjective experience can be. We refer to the dream origin of the discovery of the *humoral* (i.e., chemical) *transmission of nerve impulses* [6] in 1921 by Otto Loewi (1873–1961) of Graz, Austria [7]. Loewi himself has published two concordant accounts of these events [8a, b]. Both of them describe how he awoke in the early morning hours with a clear idea for a crucial experiment to solve a fundamental problem of neurophysiology[1] – a problem, incidentally, in which he was not consciously interested at that time.[2] He jotted down the brilliant thought and went back to sleep. The next day he remembered that he had had a most important idea the night before but was quite unable to recall during that 'most desperate day in my whole scientific life' [8] what that idea had been or to decipher the hasty notes he had made. The idea did, however, recur the next night; this time, Loewi did not take any risk, got up at once in the small hours of the morning, walked the few blocks to his Institute, and performed the 'simple experiment' [8b]. When his co-workers arrived at 8:00 a.m., the normal starting time, the work had been successfully completed.

As in Kekulé's case, the story of his dream was committed to print by Loewi himself only years later,[3] but this cannot be an obstacle to its acceptance: a background incident of this kind would have been considered out of place in a sober, factual account of serious scientific research in Kekulé's or Loewi's time just as much as in our own. Can anyone visualize getting such a story past the referee of any primary journal? However, Loewi was very fond of telling about these events, and he willingly complied with the frequent requests to do so. One of us[4] has often heard the story from him during a period of more than 20 years (1938–1961) and can testify that it was always told in essentially the same way, without major variations. The same fact is stressed by Loewi's lifelong friend, Henry H. Dale, in his memorial tribute to Loewi [9]. Even more significant, Loewi's daugh-

1 This brief note in a chemical journal is hardly the place for a detailed explanation of the problem of neurophysiology that was solved through the experiment suggested to Loewi in his dream. Briefly, the question answered had to do with the mechanism by which the two branches of the so-called vegetative nervous system: the vagus and the sympathicus, respectively, act – often in mutually antagonistic ways – upon the organs they influence. The heartbeat, for instance, is slowed down by action of the vagus, sped up by that of the sympathicus. These effects had long been thought to occur through electric currents, but Loewi's experiment, using frogs' hearts, proved that release of chemical substances is actually involved: acetylcholine from the vagus, adrenaline from the sympathicus, at least in the frog.

2 Interesting comments on this point are given in Loewi's last purely scientific paper [12].

3 Actually, an account of Loewi's dream has appeared in print in a book by Cannon [13] several years before the publication of ref. 8a. Cannon, a close, admired [8b] friend, must have heard the story being told by Loewi; he cannot have failed to secure permission for including it in his book.

4 Ullrich Weiss.

5 Anna Weiss.

ter[5] clearly recalls to this day how her father told her about his dreams and their consequences on the day after these events. She also adds a piece of information often repeated verbally by Loewi later on but omitted from his published account (presumably) as not being significant enough to get into print): that his co-workers, when they arrived in the laboratory that day, predicted that that morning's results would get him a Nobel Prize. To his daughter, he added: 'Wouldn't that be something?' The prediction was fulfilled when Loewi and Dale shared a Nobel Prize in Medicine or Physiology in 1936.

To reassure those who insist that hard work in the laboratory gathering data, with subsequent interpretation, is the only possible, indeed the only permissible way to scientific progress, we may add that for Loewi and his associates the years 1921–1938 were filled with intense experimental work to secure the results of the first experiment, to explore its further consequences, to identify the chemical agents that had been proven to be involved,[1] and to develop the field of neurobiochemistry based on these results. The enormous importance of this field is too well known to need special comment. This work was documented in a series of 14 major papers, until nonscientific events put an end to these studies.

We feel that the story of the background of Loewi's discovery proves quite conclusively that ideas for highly significant scientific research can indeed appear during sleep, as Kekulé had related. The doubts raised against such a possibility are not valid. The veracity of the great founder of structural organic chemistry need not be questioned. His delay in getting his story into print cannot possibly be held against him. Cases of this kind do not seem to be all that exceptional. While writing this note, we came across a number of examples, including a statement by Hermann von Helmholtz [10] that fruitful ideas 'often came in the morning upon waking, as Gauss also noted'. We feel that the story to which we wish to call the attention of our fellow chemists proves the validity and wisdom of Kekulé's often-quoted remark [11]: 'Let us dream, Gentlemen, then perhaps we shall find the truth, but let us beware of publishing our dreams till they have been tested by the waking understanding.'

References

1 See the account of the controversy by Seltzer, R.J. *Chem. Eng. News* 1985, (Nov. 4), 22 and literature quoted there.
2 Ramsay, O.B.; Rocke, A. *J. Chem. Brit.* 1984, 20, 1093.
3 (a) Wotiz, J.H.; Rudovsky, S. *Chem. Brit.* 1984, 20, 720; (b) *Chem. Eng. News* 1986, (Jan. 20), 3.
4 (a) Vanderbilt, B.M. *J. Chem. Educ.* 1975, 52; (b) *Chem. Eng. News* 1986, (Feb. 24), 4.
5 For an exception, see Brown, R.A.; Luckcock, R.G. *J. Chem. Educ.* 1978, 5, 694.
6 Loewi, O. *Pflüger's Archiv f. d. gesamte Physiol.* 1921, 189, 239.
7 An interesting account of Loewi's life, personality, and research is given in: Lembeck, F.; Giere, W. *Otto Loewi, Ein Lebensbild in Dokumenten;* Springer-Verlag: Berlin, 1968.

8 (a) Loewi, O. *From the Workshop of Discovery*, Porter Lectures, Series 19, University of Kansas: Lawrence, KS, 1953; p 20. (b) Loewi O. *An Autobiographic Sketch, Perspect. Biol. Med.* 1960, 4, 3.
9 Dale, H.H. Biol. *Mem. Fell. Roy. Soc.* 1962, 8, 67–89.
10 Segrè, E. *From Falling Bodies to Radio Waves;* W.H. Freeman: New York, 1985; p 224; Cf. p 61 in Cannon, W. B. in ref. 13.
11 Schultz, G. *Ber. deut. Chem. Ges.* 1890, 23, 1265.
12 Loewi, O. *J. Mt. Sinai Hosp.* 1957, 24, 1014.
13 Cannon, W.B. *The Way of an Investigator;* Norton: New York, 1945; pp 60–61.

THE LOEWI FAMILY'S WAY TO THE NEW WORLD

The life of the Loewi family in Graz was not different from that of any professor's family at the time. They lived in a beautiful villa close to the university, which had been built by an architect from Prague and was financed by Guida Loewi's father who was Professor of Chemistry in Prague and Vienna.

As common at the time, there was a cook, a maid, a governess for the children and a housekeeper and his wife taking care of the garden, the laundry and the ironing. Social events such as private concerts held at their home and dinners were arranged by Mrs. Loewi and carefully recorded in a guest book. There was no car, and therefore only short walks were made in the close surroundings of Graz. In other words, the life followed old traditions, in spite of the increasing political and economic tension in Austria at the time. Otto Loewi never mentioned the two bloody riots in 1934, one by the Socialists, the other one by the Nazis. He ignored politics really; when a young coworker, Hagen, confidentially asked Loewi shortly before 1938 whether he was not concerned about the threatening takeover of political power by the Nazis, as it had already happened in Germany in 1933, Loewi replied: 'What can they do with me? I am 64, they could perhaps force me to retire a bit earlier.'

All this ended dramatically in the night of March 12, 1938. As described in details by Otto Loewi in his Autobiographic Sketch (see pp. 11–29), by Dale (see pp. 30–53) and Brücke (pp. 54–67), he and his sons Victor and Guido were taken into protective custody and confined in the Police Prison in Graz, along with other well-known non-Jewish political figures of the city, including the Governor of Styria, Mr. Stepan, the head of the Vaterländische Front, prominent lawyers and others. Loewi shared a 'cell' with his two sons. They were not subjected to questioning. Visitors were not allowed. But the cook of the prison had formerly been the cook of the Loewi family; she went to Mrs. Loewi offering her help. She brought textbooks for Victor, which enabled him to prepare for his final medical exams. After his release, he took these exams in medicine and graduated in June 1938.

While in prison, Otto Loewi had a severe renal colic and was transferred to the clinic, where he was treated by Professor Berger, a colleague at his faculty (see p. 69). After a few days, he was returned to prison.

On the day of his release, May 6, Loewi had to sign a statement that he would leave the country within two months, and thereafter he was called to a high-ranking German Gestapo officer. One can image him feeling dreadful. 'There is someone to take you away from here', was the welcoming remark of the officer. He turned around to remove a curtain at the back of the room – there was Mrs. Loewi waiting to take him home. These episodes are not written to bring some light into these dark weeks. In my view, the subsequent steps taken by the bureaucratic machinery really revealed their cold cruelty. It was announced in official letters that he was no longer allowed to enter his department at the university, that his salary would no longer be paid and so on [for details, see Lembeck and Giere, 1968].

Otto Loewi left Graz for Berlin on September 28 and flew from there to London. Further details, including the invisible support by Sir Henry Dale, are found in the Autobiographic Sketch (pp. 11–29). He arrived in New York on June 1, 1940.

Guida Loewi-Goldschmiedt (1888–1958)

Guida Loewi had to remain in Graz. Not only did the Nobel Prize have to be transferred to the Reichsbank, but also the property of her family in Trieste, and this took some time. At the Crystal Night when, as everywhere else, the Synagogue in Graz was burned down, she left by taxi for Vienna where she had to wait for permission to leave Austria. Finally she travelled in the famous 'sealed train' from Berlin to Lisbon and arrived by boat in New York in February 1941.

Harold Low (Hans Loewi, 1909–2004)

Upon completion of the gymnasium, Harold studied law at the University of Innsbruck where he graduated. Subsequently he became a skiing instructor and glider pilot. In March 1938 he held a job in Zagreb, Yugoslavia. At the beginning of the war, he went to Peru where he worked in a copper mine in the Andes. In the early 1940s he went to the USA, joined the army and served in Europe. Thereafter, he returned to the USA and initially got involved in photography. Later on, he moved to a small town in Upper New York State, where he pursued his skiing interests.

Anna Weiss-Loewi (1911–2000)

The dramatic escape of Anna and Ulrich is documented in a TV interview, part of the ICCGW's Holocaust Oral History Project. The transcript of the interview covers 67 single-space typed pages and was corrected by Margaret, the daughter of Anna and Ulrich. Only a few points are mentioned in the following.

Her flight started in 1938 and ended with her arrival in New York in June 1941. After having read the transcript, I was deeply impressed by her precise memory of events, and even more by her unique character of never losing courage, never becoming disheartened, never complaining under the worst possible conditions, always optimistic and taking the initiative. Consider, a couple with a small baby, each carrying a rucksack and a hand bag and having absolutely no money. They moved from a village in Bohemia to Prague, to Belgium, to England, escaping during the war from Paris to a village in the Auverne, getting access to a desolate cargo ship and sailing from Marseille to Martinique, being admitted there to an internment camp and finally reaching New York. This was a long period of time full of dangers, difficulties to get the deposit for the visa and boat transfer, and the stress of getting the visas in time. Neither was the following period of time anything but smooth, which is illustrated by her first job for 7 dollars a week. In 1945, the second daughter, Margaret, was born. All the time, and even after Ulrich Weiss had joined the National Institute of Health (NIH) in Bethesda, Anna worked and finally, at a time when others retire, she became indexer at the library of the NIH. I would like to give two examples of her outstanding helpfulness: She allowed me to make a bronze copy of the bust of her father, a work of Dr. Stein, for the Graz University. When I asked her to accept a young coworker of mine for a few days at the beginning of his fellowship at the NIH, she agreed – and he was virtually adopted as her son for the next two years.

Victor Loewi (1913–1993)

Victor was accepted for training by the Sandoz Company in Basel, Switzerland, and was then sent to their office in Buenos Aires where he worked for the rest of his life. The whole matter had to be arranged in a very confidential way. My impression is that Professor Ernst Rothlin, the Research Director of Sandoz, a good friend of Dale and Loewi, sponsored the arrangements.

Geoffrey William Low
(Guido Loewi, born in 1915)

Various stations of Guido's life, filed in the *Supplement 1999*, are based on an interview by Alois Kernbauer. Here I mention only some episodes. Guido, after having graduated from the gymnasium in Graz, joined the Austrian Army as volunteer for one year. At the beginning of 1938 he spent a few months in a textile factory in Yugoslavia for training. On March 12, he heard on the radio that the Austrian Government called all military reservists to be on immediate stand-by. He took his motorcycle and drove to Graz on March 13. On that evening his parents had been listening to the dramatic speech by Chancellor Schuschnig, ending with the words: 'God save Austria!'; afterwards they played bridge. A few hours later Guido, together with his father and Victor, was taken to prison. After his release from prison, a distant relative of his father in England sponsored him and Victor as refugees. He arranged a place at a language school for him; in the family he played the role of a 'chevalier servant'. The priest in a small village in Buckinghamshire arranged a job for him in a car maintenance station outside London. One of the customers – he was the Deputy Head of Scotland Yard – was extremely helpful. When Guido expressed interest in a job within the police, he inquired at the Police Commissioners in Australia, South Africa and the Royal Canadian Mountain Police. But completely unexpectedly, he received a cable from Canada informing him that not only the entrance permit but also a paid ticket for the boat trip and a job in the laboratory of C.H. Best were waiting for him. Who was C.H. Best? His father, at that time in Brussels, had to explain to him that it was the close coworker of Sir Frederick Banting in the discovery of insulin. There is no doubt that Sir Henry Dale had been instrumental in the arrangements for Guido. With the support of Sir Frederick, Guido was soon accepted by the Royal Air Force for training as a navigator. Upon completion he was transferred to the UK. For safety reasons, as a member of the aircrew, he changed his name to Geoffrey Low. Subsequently he spent four years flying in the Far East as Flight Lieutenant with the RAF. His postwar career is mentioned only very briefly. After running a farm in England, he joined one of the large American Construction Companies as Safety Officer, and went to one of their Oil Drilling Projects in South East Asia. After his retirement he joined the US Embassy in Bangkok and took care of refugees from Vietnam and Cambodia. He became so much adapted to Thailand that he retired there, living in a small village in the south of the country.

One of the highlights in my life was the hundred-year birthday celebration of Otto Loewi by the University of Graz (F.L.). Ulf von Euler was the Honorary Lecturer. The violinist Wolfgang Schneiderhan, a friend of the Loewi family since prewar times, presented the musical contribution. Victor and his wife came from Buenos Aires, Ulrich Weiss and his wife Anna from Bethesda and Geoffrey from Bangkok! Yearly visits of Anna and Geoffrey to Graz, once also of Harold, followed.

I am deeply indebted to Geoffrey for having given me in 1962 all collected family documents which enabled Wolfgang Giere and me to condense them in the *Curriculum in Documents* published in 1968; afterwards the compiled and sorted collection of documents was transferred to the Royal Society, London. Margaret commented on the transcripts of the interviews of Anna and Ulrich (see above) in words which also hold for the interview of Guido: 'Unlike most other such narratives, which chronicle life in the Nazi concentration camps, my family managed to avoid them, this project details life on the run.'

Reference

Lembeck F, Giere W: Otto Loewi: Ein Lebensbild in Dokumenten. Berlin, Springer, 1968.

COMMENT ON THE FILM: NOBELPREISTRÄGER OTTO LOEWI

Production: F. Lembeck, W. Stoitzner (1960)
Copyright: Österreichische Phonothek, Vienna.
VHS, English version available

The film was produced in the years before 1960. It includes some of the experiments performed by Otto Loewi and an interview with him in German and English, conducted at his home in New York.

The original equipment, with some minor modifications, was used in the part of the film on his experiments. The film team came from Vienna to Graz only for one day. For illumination, they used large spotlights. A lot of time was necessary for adjusting aperture and focusing. One technician was doing one frog heart preparation after another, as the frog heart only survived for a few minutes because of the heat of the illumination. The effect of atropine was finally not included because of a lack of time. The effect of physostigmine was illustrated in a cartoon showing the use of a pipette from which the solution of an indicator was released into a buffer solution. The work of the cutter was done in Vienna without giving me the opportunity to make appropriate corrections. Finally, it seemed to me to be a miracle that a more or less acceptable film was the result (F.L.).

The interview with Otto Loewi was produced by newsreel people in New York. It has the great value of being the only film document of him.

HENRY HALLETT DALE

Sir Henry Hallett Dale. Photo taken at the Honorary Doctor Award in Graz in 1954.

HENRY HALLETT DALE – INTRODUCTORY REMARKS

One outstanding event in my life was the permission to attend a meeting of the Royal Medical Society of Edinburgh in 1949 as guest (F.L.). This was arranged for me as a young research worker at Sir John Gaddum's Department by Dr. Marthe Vogt, the only lady member of the Society. All gentlemen wore tails, and some of them were dressed in kilts. I felt rather lonely in the dark blue suit I had just bought from the Fifty Shilling Tailors. Sir Henry Dale delivered a lecture on 'Medical Research as an Aim in Life' (Dale, 1949).

I met Sir Henry Dale again when he received the Honorary Doctor's Degree of the Medical Faculty of the University of Graz in 1953; Erich Knaffl-Lenz, his former secretary at the League of Nations' International Committee on Biological Standardization of Drugs, who was born in Graz, accompanied him.

The life of Dale is documented in Wilhelm Feldberg's obituary in the *British Journal of Pharmacology* (see pp. 89–99). His in every respect excellent extensive biography (about 100 pages) published in *Biographical Memoirs of Fellows of the Royal Society* is highly recommended further reading [Feldberg, 1970].

In 1958 Dale began to write what he called an *Autobiographical Note*. At the time of his death he had written over 200 pages and had covered the period up to 1920. Feldberg acknowledged to have made full use of these memoirs.

In addition, there exists an 'autobiographic sketch' of a special design, first published in 1953. On the occasion of Dale's 90th birthday on June 5, 1965, the Wellcome Foundation organized the second edition of *Adventures in Physiology with Excursions into Autopharmacology* [Dale, 1965] containing reprints of thirty of his publications, which he had selected and to each of which he had added his comments. Dale mentioned that he tried to keep an even balance between the author's fancy and what the reader is probably interested in. As regards the latter, his comments on the papers, written many years after their publication, present the wide spectrum of his research work and include many interesting facts, such as the background of the motivation for his work, accidental observations (like the discovery of oxytocin in pituitary extract) and possible pitfalls, and he demonstrates again his careful interpretation of the results obtained. He dedicated the book to T.R. Elliot 'who had so much to do with the beginning of these adventures and, long after they have ended, is still my counsellor and friend'.

Dale was Director of the Wellcome Physiological Research Institute (1904–1914), Head of the National Institute for Medical Research (1920–1942), of the Royal Institution (1942–1946), Secretary (1935–1939) and President of the Royal Society (1940–1945). In addition he was the Chairman of the Scientific Advisory Committee of the War Cabinet. He served as the Chairman of a conference of the Health Organization of the League of Nations in Geneva in 1925 where the International Standards had been established. He, therefore, had a vast amount of responsibilities far beyond his ongoing research work. In all his official positions, Dale had the unique ability to present a clear opinion and to take immediate initiatives.

In spite of these many administrative duties, Dale maintained not only an active interest in the research which was in progress at his department, but up to 1938 he personally participated in a lot of it. Every week he tried to keep some days free for experiments; if this was not possible he discussed them with his colleagues in the mornings, and Feldberg wrote that sometimes in the evenings when he returned from a meeting, he would rush into the department before going home for dinner, eager to hear about the outcome of the day's experiments.

Dale and Loewi remained lifelong friends, a friendship beginning with Loewi's visit to London in 1903, culminating in Dale's support of Loewi after his expulsion from Graz in 1938, by raising for him the first Wellcome Trust Fellowship and by his invisible guidance of the Loewi family in finding their way to the USA.

Fifty years after the Nobel Prize Award to H.H. Dale and Otto Loewi, George. B. Koelle reviewed the life of both of them (see pp. 100–107). There has been no other article on both of them. Koelle even added a cartoon on which both can be seen at the Nobel Prize Ceremony (see p. 101).

There were many differences in striking features and in appearance between Loewi and Dale. To illustrate this, an event is quoted from Brücke's article (pp. 54–67): Dale frequently corrected Loewi's English in a polite and friendly way. But Loewi once replied: 'Please, Dale, don't correct me. I do not wish to speak English correctly, I want to speak it quickly.'

It might be noted that Dale in his Memoirs never mentioned anything about his family, although Lady Dale had an open heart for his coworkers, while Loewi talked a lot about his family life. Certainly, their lives both in Graz and London were entirely different.

On the other hand, Dale and Loewi had a great deal in common. Their reflections on science and religion might serve as an example of this and will be reviewed briefly (see pp. 108–111). Today their opinions are not less valuable than at the time when they were written. Important recent experimental results in molecular biology confront scientists again with questions of ethics or religion.

References

Dale HH: Adventures in Physiology with Excursions into Autopharmacology, ed 2. London, The Wellcome Trust, 1965.

Feldberg WS: Henry Hallett Dale, 1875–1968. Biogr Mem Fellows R Soc 1970;16:77–174.

HENRY HALLETT DALE 1875–1968

W. FELDBERG

National Institute for Medical Research, Mill Hill, London
Reprinted with permission from Br. J. Pharmacol. 1969;35:1–9

On July 23, 1968, an era came to an end with the death of Sir Henry Dale, who was regarded by physiologists and pharmacologists in every part of the world with the affectionate veneration we reserve for our greatest masters. He exerted a far-reaching influence not only over physiologists and pharmacologists, but over scientists in general. Whenever he voiced an opinion or took a stand on an issue which affected science, it was the voice of the true scientist we heard. Not that he was asked or wanted to act as the spokesman of scientists; he spoke because he felt the necessity to do so. But by his intuitive judgement, his intellectual courage, and the clarity of style with which he could formulate his ideas and express his doubts, which were held and felt by his scientific colleagues, he became their conscience. That is part of the secret how, up to his death at the age of 93, he exerted such a profound influence on the scientific world, although he had ceased to be actively engaged in research over 30 years earlier.

Dale's discoveries, which have gone into medical history, are still very much alive in the sense that they influence the stream of present-day research to an astonishing degree; they contained the seed, often hidden from his contemporaries, of the progress now being made. Dale lived with his scientific colleagues until the end, and not through his past achievements alone. Impressive was the grip he had on any problem related to his previous work that turned up. Until a few years ago, his discussion remarks at meetings of the Pharmacological and the Physiological Society were most penetrating, and sometimes initiated new series of experiments to be carried out by those who had given the papers. Until the very end, his colleagues loved to visit him at the nursing home in Cambridge to which he was confined during the last years, to listen to his stimulating criticism and, the most enjoyable reward of all, to find him enthusiastic about their new observations and ideas.

His outstanding achievements in medical research and his far-reaching influence in science have found national and international recognition. He was elected Fellow of the Royal Society in 1914. In 1936 he shared the Nobel Prize with his old friend O. Loewi. In 1943 he received the Knight Grand Cross Order of the British Empire and in 1944 he was awarded the highest honour that the King could confer upon a scholar – the Order of Merit. Another Pour le Merite was bestowed upon him in 1955 by the German Bundesrepublik, and as he once pointed out, he shared the unique position of being, holder of two orders of merit with only one other person in this country, the late T.S. Eliot. Belgium honoured him in 1944 with the Grand Croix de l'ordre de la Couronne. During the war years, from 1940 to 1945, he was President of the Royal Society, and as such became Chairman of the Scientific Advisory Committee to the War Cabinet. He was President of the British Association in 1947, of the Royal Society of Medicine from 1948 to 1950, and of the British Council from 1950 to 1955. In 1959 a medal, the Dale Medal, was struck in his honour by the Society of Endocrinology, awarded annually and a lecture – the Sir Henry Dale Lecture – delivered by the recipient. In 1961 the Royal Society established the Henry Dale Research Professorship, endowed by the Wellcome Trustees, who wished to commemorate his unique services as Chairman of their Trust for twenty-two years, as well as his outstanding contributions to science and medicine in a wider context. Several years earlier, in 1956, the Wellcome Trust had shown their deep appreciation of Sir Henry Dale in a more personal way by naming a boat 'Lady Dale,' a boat the Trust had given to the Medical Research Council Unit at Fajara, and that now cruises on the Gambia River.

Long is the list of honorary membership of foreign societies and academies, and more than twenty universities have awarded him honorary degress. It was said during the war – Dale receives new honours with each new ration book!

Dale began his scientific career immediately after having obtained a First Class in the Natural Sciences Tripos in 1898. He stayed in Cambridge for another two years before he took his clinical course at St. Bartholomew's Hospital, London, and worked in the Physiological Laboratory under Langley. Here he came under the stimulating influence of Gaskell and Anderson, and in 1900, at the age of 24, published his first scientific paper in the 'Journal of Physiology'. His main discoveries, however, were to come later and extended over a period of thirty years, from 1906 to 1936. The first eight years of these creative three decades Dale worked in the Wellcome Research Laboratories and afterwards in the laboratories of the Medical Research Council, or, as it was called until 1920, the Medical Research Committee, first as a member of the scientific staff and from 1928 onwards as the Director of the National Institute for Medical Research, which became famous under his leadership.

In 1952, when Dale was asked to make a selection for republication of his papers, which at that time exceeded 250, he chose thirty, of which twenty-six were publications of his main experimental work during these thirty years, and four were lectures given between 1920 and 1937 in which he reviewed this work. To the collection he gave the revealing title 'Adventures in Physiology,' and through arrangements with the Wellcome Trustees and the Board of Directors of the Wellcome Foundation, the volume was reprinted in 1965 to mark the occasion of Sir Henry's ninetieth birthday.

His 'Adventures in Physiology' give us not only a clear picture of his scientific achievements, but through the introduction he wrote and through the comments he made to each paper, he succeeded in reviving the past. We are given the rare pleasure of taking part in the excitement surrounding his experiments, we can admire how discoveries logically follow each other, how they reach into the future, were taken up and developed to act in the course of time as the solid foundation on which much of present research is still building.

Dale began his research at the Wellcome Laboratories with ergot, and the fungus ergot turns up again and again in his research. By discovering, that extracts of ergot reversed the pressor effect of adrenaline and sympathetic nerve stimulation, Dale introduced the first adrenaline blocking agent into pharmacology. In his 'Adventures in Physiology' he stresses the role played by lucky accidents in his research, and his ability to recognize the chances provided by such lucky accidents. These played a decisive role also in the discovery of this new action of ergot. He had been asked to determine the adrenaline content of a sample of dried adrenal glands delivered to him by the Burroughs Wellcome factory. By chance he tested it, at the end of an experiment, on a spinal cat on which he had studied the actions of ergot. To his surprise the sample when injected did not raise but lowered the arterial blood pressure, and, as he said, 'with the confidence of inexperience I condemned the sample without hesitation'. But when by 'an almost incredibly fortunate coincidence the same sequence of events was repeated in detail a week later' his interest was alerted with the result that he discovered the adrenaline reversing action of ergot.

From an accidental observation made in the course of these experiments on ergot stems another of Dale's important discoveries, that of the oxytocic action of pituitary extract. He had used such an extract in one experiment as a control, and it happened that in this experiment he was also recording the tone and rhythm of the cat's uterus, and thus 'by chance' as he said, observed the potent stimulating action on the uterus muscle, a discovery which probably led to the most frequent application in practical medicine of pituitary extract.

Another example of a fundamental discovery which, as Dale tells us, 'arose from one of my most fortunate accidents', was his demonstration that the anaphylactic contraction of smooth muscle resulted from the formation of cell-fixed antibodies. Whilst studying the effects of horse serum and comparing them with those of histamine on the isolated guinea-pig uterus, he encountered one preparation which responded with a maximal contraction to the application of the serum in a usually inactive dilution. He made enquiries about the guinea-pig because he knew that, at that time, the Wellcome Laboratories were using guinea-pigs for the assay of antitoxins which involved many small injections of horse serum. His suspicions were confirmed: he had been supplied with a guinea-pig used for these assays. The results of his subsequent experiments which were initiated by this 'most fortunate accident' have changed our conception of anaphylaxis and have given us an understanding of the relation between hypersensitivity and immunity. As Dale expressed it, anaphylaxis results from a predominantly cellular fixation of an antibody which, when it circulates in excess, is recognized as the cause of immunity.

The main topics of Dale's other research were concerned with the physiology and pharmacology of sympathomimetic amines, histamine and acetylcholine. In each instance the work began with the identification of the substance in extracts of ergot; of tyramine and histamine, in 1910 together with G. Barger, and of acetylcholine three years later as the result of an accidental observation – 'the lucky encounter with a peculiar ergot extract'. The extract, which had been sent to him for routine examination, had aroused Dale's interest because of its unusual activity; it caused profound inhibition of the heart beat and had other actions resembling those of muscarine. He secured the whole batch of the extract for further investigation by his chemist colleague, A.J. Ewins, who succeeded in obtaining a few milligrams of the abnormal active constituent as a platinum salt. Its properties however, were not those of the stable muscarine but rather those of a very labile ester, and there seemed little hope of identifying the substance until Dale, as he said, 'recalled into conscious memory an observation made some eight years earlier' by Hunt and Taveau on the enormous potency of the unstable ester acetylcholine. When Ewins accordingly made Dale some acetylcholine for comparison, its identity with the substance from the ergot extract was put beyond doubt. Fifteen years later, in 1929, and again as the result of an 'incidental' and 'quite unexpected' observation whilst studying the distribution of histamine in animal tissue, Dale, together with H.W. Dudley, isolated acetylcholine in extracts of horse spleen and provided the first direct proof for its occurrence in tissues of the animal body.

The discovery of tyramine in ergot was in part responsible for the research he and Barger carried out, in 1910, on the relation between chemical structure and pharmacological action of a number of amines related to adrenaline. For these amines he coined the term 'sympathomimetic amines' now in general use. In later years this paper, which is a mine of information, gave Dale some cause for regret because it 'shows evidence of missed opportunities' as he expressed it. Having found that the N-propyl members of the amino-aceto-catechol and of the amino-ethanol-catechol series had a weaker action than their lower homologues, they did not examine the N-isopropyl member of the amino-ethanol-catechol and thus missed the discovery of the therapeutically valuable isoprenaline, with its strong inhibitory actions. Further, having pointed out that adrenaline, which only incompletely reproduced sympathetic actions, could not be the transmitter of sympathetic nerve impulses, they discussed but discarded the possibility of noradrenaline being the main transmitter because at that time noradrenaline was nothing but a synthetic curiosity and was not known to occur in the body. We admire the caution, but it is frustrating to read the few sentences in the discussion dealing with this problem and to realize how very near he was to the correct solution. 'I failed to jump to the truth, and I can hardly claim credit for having crawled so near and then stopped short of it' he says, to which we may reply, with Molière, 'c'est le privilège d'un grand homme de faire impunément des défauts.'

From the discovery of histamine and acetylcholine in ergot arose the study of their pharmacological actions. Our present knowledge of the actions of these two potent substances is mainly derived from these studies which were published between 1910 and 1920. They form some of the finest pharmacological analyses ever made, particularly those of the circulatory effects. They have become models for the analysis of drug actions, and we admire the simplicity of the methods Dale used to reveal the mechanisms of their actions.

The pharmacological studies of histamine which were begun with P.P. Laidlaw brought to light nearly all its main actions, those on plain muscle, on capillaries, gland cells, the suprarenal medulla and on the central nervous system; the one important action missed, because they did not look for it, was the action on the oxyntic cells of the gastric mucosa which results in strong acid gastric secretion.

Having shown that the fall in arterial blood pressure produced in cats and dogs by small doses of histamine resulted from vasodilatation, Dale, together with A.N. Richards and later with J.H. Burn, analysed this effect and was able to show where in the vascular tree histamine produced its dilator effect. Simply by observing the changes in skin colour of unpigmented pads of the feet in cats, and by measuring in a limb the volume changes during the fall in blood pressure, or in perfusion experiments during the changes in outflow, he was able to distinguish between the dilatation of capillaries and that of arteries and arterioles. From the results obtained with these methods he could conclude that in cats histamine dilated the capillaries in contrast to acetylcholine, which dilated the arteries, but that in dogs, histamine dilated capillaries and arterioles. But histamine had a vasoconstrictor action as well, which was revealed particularly in perfusion experiments. In later writings Dale has summarised the position in different species and pictured 'the action of histamine on vascular tone, in those species in which it has been possible to analyse its incidence, as changing in every case, but at different levels in the vascular branching from a more central arterio-constrictor to a more peripheral dilator effect; and this change occurs at levels moving towards the periphery, as the species are considered in the following order – monkey (and presumably man), dog, cat and, finally, rabbit.' Time has not changed the validity of this statement.

With the knowledge of a predominantly capillary site of the vasodilator action of histamine, Dale and Laidlaw, in 1918, studied the mechanism of histamine shock. This study was 'undertaken as a contribution to a co-operative war study of conditions of varied origin, then being included in a general description as "secondary wound shock"-conditions apparently resulting in most cases from haemorrhage, exposure, infection and toxaemia in unknown proportions.' They hoped that a study of the artificially reproducible histamine shock might indicate by analogy some common factor in the resultant syndrome and suggest some appropriate treatment.

Using the simple method of measuring the corpuscular content of blood with the haematocrit, Dale and Laidlaw showed that large doses of histamine given intravenously, particularly by slow infusion, not only dilated the capillaries but also caused a general increase in their permeability, thereby allowing plasma to escape from the vessels into the tissues as seen by the striking increase in corpuscular content of blood withdrawn when the shock had fully developed. The ensuing reduction in circulating blood volume and the tendency of the remaining blood with its increased viscosity to stagnate in the dilated capillaries and venules instead of returning to the heart were the characteristic features of histamine shock. These findings emphasized 'a deficiency in the volume of blood in effective currency, as a common factor in conditions of so-called secondary wound shock and pointed to the addition, by transfusion to the circulating blood of a fluid which the morbidly-permeable capillaries might retain – blood or plasma – as the most immediately effective remedy.'

To Dale we also owe our knowledge of the occurrence of histamine as a natural constituent in many animal tissues, as was convincingly shown in 1929, when he, together with C.H. Best, H.W. Dudley and W.V. Thorpe, isolated histamine from alcoholic extracts of fresh liver and lung, and together with Dudley from alcoholic extracts of spleen, and in amounts sufficient to account for the histamine-like activity of such extracts. Once again, circumstances of an accidental kind had provided the immediate stimulus for this study. A liver extract had been acquiring some reputation in Canada for high blood pressure, and Dale, as he tells us, was asked to examine this extract to verify and standardize its activity, and if possible to find out the substance responsible for it.

In 1918, however, histamine was not known to be a natural constituent of animal tissue, and Dale and Laidlaw not only avoided any premature assumption of a direct participation of histamine in traumatic shock or in the local vascular reactions of the tissues to injury or irritation, but as he later wrote, 'I have been quite shocked on reading anew one passage of our discussion to discover how we seem to have gone almost out of our way even to preclude this possibility from consideration.' Naturally, Dale was aware of this possibility, but to be cautious in making generalised conclusions based on analogy alone was characteristic of him. In this attitude, this ability to be able to wait, we recognize the true scientist who humbly pays his respects to nature's secrets. His attitude may serve as a warning to those of us – for whom it does not come too late – to be cautious when speculating about the physiological significance of newly discovered facts and not to try to prove more than they warrant. Dale never erred in this direction and thus failed to experience what Thomas Huxley described as 'the great tragedy of science – the slaying of a beautiful hypothesis by an ugly fact.'

Even after his discovery of histamine in animal tissues, when discussing its function for the symptoms of cell injury or irritation, of anaphylaxis or of shock-like conditions, he emphasized not the similarities but the discrepancies between these symptoms and those produced by histamine, and did not attribute to histamine a role greater than it could possibly play. As Adrian once said about Dale, he has been more concerned to apply the brake than to be first in the gold rush, but the gold he has found will keep its value. Dale's critical discussion of the role of histamine in physiological and pathological reactions given in his Croonian Lectures of 1929 has not diminished in significance, his arguments are as pertinent today as they were forty years ago.

In 1914, when Dale published his experiments on the action of some esters and ethers of choline, including acetylcholine, he could scarcely have foretold that in the course of time these investigations would lead to a revolutionary change in our concept of neuromuscular and synaptic transmission. In this paper Dale did not describe just this or that action of these derivatives of choline, but by recognizing that they shared some of their actions with muscarine, others with nicotine, he established the dual action of acetylcholine. It acted like muscarine on smooth muscles, heart and gland cells, and these muscarinic actions were abolished by atropine. But Dale showed that acetylcholine acted also like nicotine, in that it activated or excited the cells of the adrenal medulla and of the sympathetic ganglia and, as shown later, that it depolarized the motor endplates of striated muscle; and these nicotinic actions were abolished by large paralysing doses of nicotine. The real importance of the dual action of acetylcholine, that is, its physiological implication, became evident only about twenty years later, when Dale and his colleagues demonstrated the role of acetylcholine in the transmission of nerve effects. But already in 1914 Dale pointed out that if there were any evidence for the presence of acetylcholine in animal tissues – the evidence was later supplied by him and Dudley in 1929, acetylcholine would be a most suitable transmitter substance of nerve effects at certain divisions of the autonomic nervous system. He also suggested that the evanescence of the acetylcholine effects might be due to the activity of an esterase, thus he predicted the presence of cholinesterase in the animal body sixteen years before it was discovered independently in two places: by Matthes in Dale's laboratory, and by Engelhardt and Loewi in Austria.

The publications concerning the role of acetylcholine as transmitter of nerve effects that appeared between 1934 and 1937 from the National Institute for Medical Research are based on Dale's original discovery of the dual action of acetylcholine; they all, whether he is one of the authors or not, bear his stamp. Those who took part in these exciting investigations were G.I. Brown, J.H. Gaddum, Marthe Vogt and myself.

Acetylcholine was shown to be the transmitter not only of postganglionic parasympathetic but also of some postganglionic sympathetic fibres, for instance, of those innervating the sweat glands in cats. Thus the transmitter function of acetylcholine was not limited to one of the two anatomical divisions, parasympathetic or sympathetic, of the autonomic nervous system. However, the discovery of the nicotinic transmitter function of acetylcholine entailed a much wider implication – it came as a real shock. The finding that acetylcholine was the neuromuscular transmitter at the motor endplate of striated muscle and the synaptic transmitter in a sympathetic ganglion went against all the, often ingenious, ideas which had been propounded up to that time by those working in this field because it had been taken for granted that these transmission processes were physical – purely electrical events. How many today realize or recall the complete change in concept that took place as a result of the discovery of the nicotinic transmitter function of acetylcholine.

When these discoveries were made, Dale at once realized the necessity, in order 'to promote clear ideas', of a terminology which would distinguish nerve fibres not with regard to their anatomical origin, but with regard to their chemical transmission, due in one case to acetylcholine, in the other to a substance like adrenaline, now known to be noradrenaline. Therefore as early as 1933 he coined the terms 'cholinergic' and 'adrenergic' for use in this sense and to be applied to nerve fibres, neurones or transmission processes. Our present knowledge of the distribution of cholinergic and adrenergic neurones in the peripheral nervous system is based on the results obtained in the thirties by Dale and his associates.

Very soon Dale had to defend his new terminology when the expressions 'cholinergic' and 'adrenergic' were used not only to designate neurones but substances as well. In 1935 Dale protested and wrote: 'The application of the term cholinergic to a substance supposed to act like acetylcholine threatens a loss of precision in the use of these words and therewith of their value.' However, the misuse continued and in 1953 Dale implored the writers in this field to resist the impulse to widen the application of these terms by allusion or 'transference of epithet.' 'This can be attractive in a poem, no doubt, but for a scientific term it is destruction of its only value, precision.' And in 1954 Dale again complained that he suffered, as he expressed it 'from the actions of colleagues in kidnapping my verbal offspring for what I regard as improper uses.' Later, Dale realized the need for an additional terminology, one which would denote the sensitivity of a structure to the transmitter substances, and he suggested the terms 'cholinoceptive' and 'adrenoceptive.'

Dale had not only to defend his terminology, but also the theory of the transmitter functions of acetylcholine, and we again admire the pungency of his style when refuting the suggestion of some authors that acetylcholine was the transmitter not only for synaptic transmission, but also for the propagated nerve impulse along the nerve. 'Such a conception,' he wrote, 'offers of course the kind of intellectual satisfaction which many minds obtain from a simplifying and co-ordinating generalization.' 'But,' he continued, 'the ingenuity of its supporters is sorely taxed to discover even plausible ways of escape from the facts which contradict it.'

One other major contribution to science and particularly to therapeutics for which humanity owes a great debt to Dale must not be forgotten. It concerns his work on the standardization of drugs, of posterior pituitary extract, neoarsphenamine, insulin and digitalis. He showed that it is essential to use a common standard reference preparation when estimating the potency of a preparation containing one of these substances, and he was able, as the chairman at a conference of the Health Organisation of the League of Nations at Geneva in 1925, to persuade the other members to adopt international standards for these substances. The benefit of this arrangement became so obvious that at later conferences international standards were adopted almost as a matter of course for a number of other substances. 'How great a contribution this has been to therapeutics throughout the world,' writes J.H. Burn on the occasion of Dale's eightieth birthday, 'it is quite impossible to indicate or measure. It is very largely due to Dale's skilful persuasion that instead of confusion there is order and co-operation in a field where national prejudices are sometimes just as capable of preventing agreement as in politics generally.'

Of the many great scientists Dale met during his life there were two he particularly admired: Rutherford and Ehrlich. About Lord Rutherford, Dale has left us 'Some personal memories' in a lecture given in New Zealand and republished in the delightful little book 'An Autumn Gleaning', a selection made by him in 1954 of his lectures and addresses. For a time Dale even contemplated writing a biography of Rutherford, writing it so, as he said, that a schoolboy of 14 would enjoy it. Ehrlich he had met when, as a young man, he had gone to Frankfurt to work in his Institute. He wanted to make a more direct contact with Ehrlich's way of thinking, but the work he was asked to do did not appeal to him. So he stayed for only four months, but as he often pointed out, he was glad to have seen something of this special genius who fascinated him during his whole life. After the last war Dale was responsible for the publication in his country of Ehrlich's collected papers. They make up three large volumes compiled and edited by Dr. F. Himmelweit, but under the editorial direction of Sir Henry, who wrote an introduction to each volume. These introductions, particularly the first one, reveal Dale as a fine historian, and rank as outstanding contributions to the history of medical science. Here a great scientist pays homage to the greatness of another genius of a different kind. Because different they were! And yet what Dale wrote about Ehrlich's way of working is a true reflection of Dale's own approach to research. This is perhaps not surprising, for psychology works in both directions, exposing the biographer as well. When Dale writes of Ehrlich's readiness, on occasion, to talk freely about what he regarded as his own special methods and experiences in research, and how Ehrlich at times thought of himself as essentially an opportunist, owing his successes mainly to a quick recognition of chances presented by lucky accidents of observation, do we not recognize in this description nearly the same vocabulary Dale liked to use, and with such relish, when speaking of his own successes? 'By one of my greatest strokes of good fortune,' said he, referring to his first attempt in practical pharmacology – which led to the discovery of the adrenaline reversal by ergotoxin – 'it was to give me an immediate opportunity of making a mistake of my own – a really shocking "howler"!' Dale was endowed with the same quasi-intuitive subconscious reasoning that he attributed to Ehrlich. Of Dale we can say what he said about Ehrlich when quoting Helmholtz, who exclaimed in admiration of Faraday: 'He smells out the truth.' Those who have worked with Dale were struck by just this ability of his, and many have testified to it with submissive admiration.

Full justice to Sir Henry Dale's personality and deep humanity would not be done without recalling his attitude to the grave problems which confront the scientist of today. He knew that relentless vigilance is required to prevent any undue curtailment of the freedom of science. We are grateful to him for having defended that freedom whenever and wherever in the world it was threatened. Let us again listen to some of his warnings, for instance those he expressed during the war in his Presidential address to the Royal Society in 1941. 'I see danger,' he said, 'if the name of science, or the very cause of its freedom, should become involved as a battle cry in a campaign on behalf of any political system, whether its opponents would describe it as revolutionary or reactionary.' And, he went on, 'This Society, with its firm and unbroken tradition of complete aloofness from political controversy, may still find it an important part of its function to keep watch and, if necessary, to stand without compromise, for the right and the duty of science to seek the truth for its own sake, in complete freedom from any kind of extraneous influence.' It was in the defence of the freedom of science that at the height of the Lysenko controversy Dale resigned from the Academy of Sciences of the USSR. This he made clear in his letter of resignation in which he expressed concern that as a result of the policy adopted – by not allowing criticism of Lysenko's conclusions and by dismissing from positions of authority those who persisted in applying scientific standards to the argument – 'The whole great fabric of exact knowledge still growing at the hands of those who have followed Mendel, Bateson and Morgan, is to be repudiated and denounced.'

We are proud of the stand Sir Henry took with regard to the atom bomb. In a review written in 1951 about a book on the second atom bomb that fell on Nagasaki he wrote, 'We begin to talk glibly, almost with wry humour, of the possibility of a hydrogen bomb which, with its yet incalculable range and power of devastation, could serve no visible or conceivable purpose of peace but only that of an immeasurable destruction or the threat of it.' He warned us that in considering a possible atomic war in merely quantitative terms we easily lose our sense of human values, 'And,' he concluded, 'nothing, it seems, but a general resurgence of human ideas can halt the world now on the ever more precipitous slope.'

And when on his ninetieth birthday he was asked in an interview by the BBC about a letter he had written to 'The Times' in August 1945, and about the hope expressed in this letter that science should become international once again after the war with Japan was over, he reiterated his creed. Let us again hear this, his final appeal, which gave voice to the uneasiness that is felt as strongly now as it was at that time by every true scientist and bears witnesses to the deep responsibility Dale felt towards science and humanity. 'We have tolerated much, and will tolerate anything to ensure victory for freedom, but when the victory has been won we shall want the freedom. I still believe that freedom of scientific knowledge among nations should be the aim of national policy. I still believe that the true spirit of science working in freedom, seeking only truth and fearing only falsehood and concealment, would still offer a lofty and austere contribution to man's moral equipment, which the world cannot afford to lose or to diminish.'

FIFTY YEARS AFTER THE NOBEL PRIZE AWARD TO H.H. DALE AND OTTO LOEWI

G.B. KOELLE

Reprinted with permission from The Johns Hopkins University Press: Perspect Biol Med 1985;28: 434–439

Henry Hallett Dale 1875–1968

In every field of human endeavour success or its absence is determined to a considerable extent by chance. This is true especially in the arts and in the entertainment professions, but it holds also for the sciences. However, a small number of individuals are so endowed with ability and perseverance that their success is assured and they inevitably rise to greatness. Such a man was Henry Hallett Dale. He was undoubtedly the outstanding pharmacologist of the present century. His achievements were recognized appropriately by his election to the presidency of the Royal Society, by his numerous degrees and distinctions (OM, GBE, MD, FRCP, FRS), knighthood, and a Nobel Prize which he shared with Otto Loewi in 1936. This year marks the fiftieth anniversary of that event.

Sir Henry Dale was also unique among biomedical scientists of his time in that throughout most of his career he never held a primary academic Post. After he received his degree in physiology from Cambridge, Dale worked there for two more years as a research fellow under J.N. Langley. He then began his clinical studies for a medical degree at St. Bartholomew's Hospital in London (where a plaque in the pathology museum marks the site where Sherlock Holmes and Doctor Watson first met). During that time he also continued his physiological investigations in the laboratory of Ernest Starling. It was there, in 1902, that he met a visitor from Germany, Otto Loewi, with whom he formed a lifelong friendship. Many years later their careers became remarkably intertwined. Dale spent the equivalent of a brief postdoctoral stay in the laboratory of Paul Ehrlich at Frankfurt am Main. Shortly afterwards in 1904 he accepted an appointment as pharmacologist at the newly established Wellcome Research Laboratory in Lon-

Dale and Loewi at the Nobel Prize Ceremony 1936 (Stockholm Newspaper Cartoon).

don, where he remained for ten years. Then, following World War I, he moved to another pioneering venture, the National Institute for Medical Research at Mill Hill, London. He subsequently became Director and remained in that capacity until his retirement in 1942. During this period, Dale was appointed an original Trustee of the Wellcome Trust, a Post which he held, eventually as Chairman, for twenty-one years.

When Dale joined the Wellcome Research Laboratory it was a unique institution, the first established by industry for the support of investigation unrelated to immediate financial returns. Sir Henry Wellcome's suggestion that Dale undertake a study of the pharmacology of ergot resulted in most of Dale's major contributions. From ergot, he and his colleagues isolated and investigated ergotoxine, pressor amines, histamine, and acetylcholine.

Ergotoxine (subsequently found to be a mixture of three alkaloids) was shown by Dale to block or reverse the pressor effect of epinephrine; thus it was the first discovered adrenergic blocking agent. In recent years, β-adrenergic blocking drugs have come to the forefront in the treatment of hypertension.

From a purified batch of ergot, Dale and his colleagues isolated a number of pressor amines. He and G. Barger then undertook a systematic investigation of the pharmacology of synthetic amines related to epinephrine. From this they concluded in 1910 that norepinephrine reproduces the effects of stimulation of sympathetic nerves more closely than epinephrine; the latter had been suggested by T.R. Elliott as a sympathetic neurohumoral transmitter a few years earlier. It was not until 1946 that von Euler in Sweden demonstrated that norepinephrine is indeed the transmitter of sympathetic fibers.

Dale's studies of the actions of histamine resulted in the demonstration of its production of capillary dilatation and increased capillary permeability. These findings were extended to elucidate the role of histamine in anaphylactic shock and local tissue injury, and the concept of fixed and circulating antibodies.

Sir Henry Dale's greatest contribution was undoubtedly his major role in establishing the function of acetylcholine as a neurohumoral transmitter. After it had been found in an extract of ergot, he and his associates embarked on an extensive exploration of its pharmacology. In a paper published in 1914, Dale distinguished between the muscarine-like and nicotine-like actions of acetylcholine, and characterized it as an ideal candidate for the role of the transmitter of parasympathetic fibers. The chief obstacle was that acetylcholine had not been found in animal tissues, but as Dale pointed out, this could have been due to its instability.

The stage was thus set for Otto Loewi's crucial experiment in 1920, when he demonstrated neurohumoral transmission from the vagus nerve to the frog heart; he subsequently showed that the transmitter was probably acetylcholine, that it was rapidly broken down by cholinesterase, and that the latter was inhibited by physostigmine.

Following the chemical identification of acetylcholine in horse spleen by Dale and Dudley in 1929, Dale and his associates at the National Institute for Medical Research (including E. Bulbring, G.L. Brown, J.H. Burn, W. Feldberg, J.H. Gaddum, F.H. MacIntosh and M. Vogt) devoted most of the subsequent decade to establishing the extensive distribution of cholinergic fibers and the biochemical and physiological details of cholinergic transmission. In the course of these studies, Dale introduced a number of terms that have since become standard: sympathomimetic and parasympathomimetic; adrenergic and cholinergic; adrenoceptive and cholinoceptive; dopamine; muscarinic and nicotinic. It is notable that the last two, introduced to describe the selective effects of acetylcholine and related drugs at peripheral sites and the corresponding receptors, remained appropriate after cholinergic transmission was demonstrated in the central nervous system.

The details of Dale's service on numerous advisory boards and committees, and his scientific contributions to his country and its allies in two World Wars can be found in Refs 3–8. It is interesting to note that his career also illustrates the importance of being in the right place at the right time. Dale's two years in the laboratory of J.H. Langley, at the turn of the century, occurred ten years following the mapping out of the sympathetic and parasympathetic nervous systems by Langley and W.L. Dickenson, through the identification of their respective ganglia by treatment with nicotine. While Dale was there, Langley was engrossed with a study of the similarity of the effects of sympathetic nerve stimulation and the injection of epinephrine (which M. Lewandowski was doing independently in Russia). This work was extended a few years later by Langley's student, T.R. Elliott, who proposed that the release of an epinephrine-like substance might mediate the actions of sympathetic nerves, as mentioned above. When Dale met Loewi in Starling's laboratory in 1902, they probably discussed the possibility that a muscarine-like substance might be a vagal transmitter (in keeping with some recent experiments of O. Schmiedeberg), for Loewi suggested this to W.M. Fletcher a few months later. Dale's identification of acetylcholine in ergot was greatly facilitated by R. Hunt's earlier demonstration that acetylation of choline (a known constituent of mammalian tissue) increased its vasodepressor potency by five magnitudes. His subsequent interactions with Loewi are well known. Their sharing of a Nobel Prize for what remains the greatest advance in neuroscience of the twentieth century was most deserved. (In this connection it might be suggested that the two greatest over-sights by Nobel Prize Committees were probably Tolstoy in literature (he was nominated ten times) and E.K. Marshall, A.N. Richards, and Homer Smith for their joint contributions to renal physiology.)

In the summer of 1964 I called on Sir Henry Dale, when he was 89, at the headquarters of the Wellcome Trust in London. At that time the Burn and Rand hypothesis (that a 'cholinergic link' is involved in adrenergic transmission) was a subject of hot debate. I liked the hypothesis, for I felt that it explained certain observations that were difficult to interpret otherwise; however, it was opposed by the majority of pharmacologists. Sir Henry asked: 'What do you think of that young fellow Burn's idea?' (J. Harold Burn was 72 at the time.) I was delighted when he nodded in agreement ('It seems like a reasonable suggestion') with my affirmative response. It brought home an interesting generalization: older people are more inclined than younger to judge favourably on the radical proposals of others.

Sir Henry Dale remained intensely interested in all areas of medical research and was readily available for enthusiastic discussion up to the time of his death in 1968. This marked, as W. Feldberg noted, the end of an era.

Otto Loewi 1873–1961

Few experiments have so greatly influenced the direction of biological thought and investigation as did the simple but ingenious one performed by Otto Loewi on Easter Monday in 1920. Two nights earlier he had awakened, jotted down a rough protocol, and returned to sleep. The following morning he was unable to decipher his scrawl. But that night the same thought recurred; this time, he transcribed it more carefully, and the next day went directly to the laboratory and conducted the experiment.

In his own words: 'The hearts of two frogs were isolated, the first with its nerves, the second without. Both hearts were attached to Straub canulas filled with a little Ringer solution. The vagus nerve of the first heart was stimulated for a few minutes. Then the Ringer solution that had been in the first heart during the stimulation of the vagus was transferred to the second heart. It slowed and its beats diminished just as if its vagus had been stimulated. Similarly, when the accelerator nerve was stimulated and the Ringer from this period transferred, the second heart speeded up and its beats increased. These results unequivocally proved that the nerves do not influence the heart directly but liberate from their terminals specific chemical substances which, in their turn, cause the well-known modifications of the function of the heart characteristic of the stimulation of its nerves'. Thus, the foundation of the theory of neurohumoral transmission was established.

Within the next few years Loewi showed that the 'Vagusstoff' is acetylcholine, and that it is rapidly broken down by the enzyme cholinesterase. He discovered also that the latter is inhibited by physostigmine. Several years later he found that the excitatory factor released by the accelerator nerve is epinephrine (the sympathetic transmitter in amphibians).

At this period, Loewi believed that neurohumoral transmission is confined to the autonomic nervous system. The extension of the concept of cholinergic transmission was accomplished largely at the National Institute for Medical Research in London, in the laboratory of Henry Hallett Dale. His collaborators in this effort included G.L. Brown, Edith Bulbring, J.H. Burn, W. Feldberg, J.H. Gaddum, F.C. MacIntosh, and Marthe Vogt. Collectively, they showed that acetylcholine is the transmitter of postganglionic parasympathetic fibers, exceptional postganglionic sympathetic fibers (e.g., to the eccrine sweat glands), preganglionic fibers in general, somatic motor fibers, and some tracts of the central nervous system.

Otto Loewi was born in 1873 in Frankfurt am Main. As a student at the local Gymnasium he was particularly involved with the classics (nine years of Latin, six of Greek) and humanities, in which he received excellent grades; his marks in physics and mathematics were generally poor. Following graduation, he wished to pursue further studies in art history, but his more practical minded family prevailed on him to enroll as a medical student at Strassburg University. Although there were a great many brilliant scientists on the medical faculty at Strassburg, Loewi soon learned that these are not necessarily inspiring teachers. Accordingly, he attended regularly only the lectures of the anatomist, Gustav Schwalbe, and spent most of his time auditing lectures in history, architecture, and philosophy.

In keeping with the custom at the time, Loewi spent his third year of medical school at the University of Munich. Here his major preoccupations were the opera, theatre, and art museums, in all of which Munich was outstanding. When he returned to Strassburg the following year, his interest in medicine was finally kindled by Bernhard Naunyn, chief of the medical clinic and a famous experimental pathologist. His other major recollection of this period was his introduction to 'On the Life of the Atlantic Salmon in Fresh Water' by Friedrich Miescher. Loewi credited this book as the factor that inspired him to enter biology and physiology, and eventually pharmacology.

Following graduation, Loewi spent brief stints in Franz Hofmeister's biochemical institute at Strassburg and in the medical clinic of von Noorden in Frankfurt. He quickly decided that his primary interest lay in basic rather than clinical medicine, and shortly thereafter he obtained an appointment with Hans Horst Meyer, the pharmacologist at Marburg an der Lahn. The ten years (1898–1908) with Meyer, first in Marburg and subsequently in Vienna, were both happy and productive. Loewi's investigations during this period covered many areas: phlorhizin diabetes, amino acid metabolism, and renal function. In 1902 he spent an extended visit in Ernest Starling's laboratory in London, where he formed lifelong friendships with a number of British and visiting scientists, including Henry Dale.

Following his return to Marburg, Loewi resumed his studies of diuretics, but his interests were evidently considerably broader. One day during a walk through the woods with Walter Fletcher, his collaborator from Cambridge, he expressed the thought that the vagus nerve might release a muscarine-like substance to produce its cardioinhibitory effect. Apparently this thought remained dormant in his mind for twenty years, while he worked in other areas, until he performed the crucial experiment described at the beginning of this article.

In 1904 Meyer was appointed to the chair of pharmacology in Vienna, where Loewi joined him the following year. Once again he was able to enjoy the benefits of the opera, the symphony orchestra, and the theatre; he formed a strong friendship with the renowned actor Joseph Kainz. His laboratory work at this time was related to carbohydrate metabolism and, at last, the autonomic nervous system. In the latter area he made the important observation that cocaine potentiates the response to stimulation of sympathetic nerves and to injected epinephrine. Although the underlying mechanism was not unravelled for more than half a century, this finding provided a means for markedly increasing the sensitivity of bioassays of endogenous catecholamines.

Two important events in Otto Loewi's life occurred in 1908–1909. He met and married Guida Goldschmidt, his constant companion and supporter for over fifty years, and he accepted the chair of pharmacology at Graz. Although Graz was the second largest city in Austria, it had only one-tenth the population of Vienna; nevertheless it also had an excellent opera, symphony orchestra, and theater, and Loewi took full advantage of these offerings. In spite of such distractions, Loewi managed to deliver five lectures per week during the winter term, a record which few people have matched. Loewi's research at Graz was centred again on carbohydrate metabolism and on the interactions of digitalis and calcium.

As soon as the significance of his critical experiment in 1920 was recognized, Loewi gained world-wide fame. He was invited to lecture and feted throughout Europe and the United States. In 1936, he and Sir Henry Dale shared the Nobel Prize in physiology. The tragic aftermath of this event has been related elsewhere. On March 12th, 1938, the day following the Anschluss, Loewi was awakened in the middle of the night by a gang of stormtroopers and dragged off to jail along with two of his sons. He was held for two months, until he could ransom himself by retrieving his share of the Nobel Prize from a Swedish bank and turning it over to the Nazi party. During this horrifying period his main concern was with some experiments he had completed the day before his arrest, showing that acetylcholine is completely absent from afferent nerve fibres. He feared that he might be killed before this finding could be communicated. However, he was eventually permitted a pencil and a postcard, with which he sent the essential data to Die Naturwissenschaften; after that he relaxed and awaited his fate.

Immediately following his release, Loewi was deprived of all his possessions and expelled from Austria. His wife was not permitted to leave for three more years, the time it took for the transfer to the Nazi of some property she held in Italy. Loewi went directly to London, then accepted an eight month visiting professorship at the Université Libre in Brussels. Shortly after his return he was invited by George Wallace to come to New York University College of Medicine as a permanent research professor. He arrived on June 1, 1940 at the age of sixty-seven and continued as an active investigator for fifteen years, until six years before his death in 1961. His summers were spent at Woods Hole, where his warmth, wisdom, and ready accessibility earned him the reputation of its outstanding celebrity.

I was privileged to visit Otto Loewi a few years after his arrival at New York University. He was enthusiastically infusing some frog hearts with calcium solutions, 'tying up loose ends'. In our conversation he related how a team from a neighbouring school had invited him to participate in some experiments with the object of demonstrating that migraine headache is caused by the release of excess acetylcholine. He said, with a twinkle: 'I considered it, but the idea seemed far-fetched. Then I thought, I have a Nobel Prize, my reputation; I can't make it any better, only worse! So I declined. They went ahead without me. They collected some urine from a patient during a migraine attack and poured it on an isolated frog rectus. What else could the poor thing do? So they published a paper.'

Otto Loewi was truly a great man. He was also a simple, kindly, and empathetic one, which all great men are not. He lived at a time when the pace of life was more leisurely than now; when one could immerse oneself in an operatic performance without the distraction of worrying about a grant application due the following week. Nevertheless, many of his precepts still hold true: The advantages of a broad classical education, to sharpen the mind to perceive relationships that might otherwise be missed. The policy of investigating in a wide variety of fields before concentrating on one that suddenly becomes ripe for a major advance. The maintenance of a keen interest in areas beyond the scope of the laboratory, to allow hiatuses during which inspiration for new experimental approaches frequently occurs. All these principles are illustrated in Otto Loewi's life and achievements.

References

1 Barger, G. and Dale, H.H. (1910–1911) J. Physiol. (London) 41, 19–59.
2 Dale, H.H. (1914–1915) J. Pharmacol. Exp. Ther. 6, 147–190.
3 Dale, H.H. (1963) Annu. Rev. Pharmacol. 3, 1–8.
4 Evans, L.B. (1985) Investment in Research, The Burroughs Wellcome Fund, Research Trangle Park, NC.
5 Feldberg, W. (1969) Br. J. Pharmacol. 35, 1–9.
6 Holmstedt, B. and Liljestrand, G. (1982) Readings in Pharmacology, Raven Press.
7 Kolle, G.B. (1985) Perspect. Biol. Med. 28, 434–439.
8 Symposium an neurohumoral transmission. Physiological Society of Philadelphia (1954) Pharmacol. Rev. 6, 1–131.
9 Loewi, O. (1960–1961) Perspect. Biol. Med. 4, 3–25.
10 Loewi, O. (1932–1933) The Harvey Lectures, Series 28, 218–233.

RELIGIOUS REFLECTIONS OF DALE AND LOEWI

Dale's Reflections

One need investigate no further than Wilhelm Feldberg's memoirs of Dale to see how critical his view on the relations between science and religion was. Dale's careful selection of words provided an outstanding representation of facts, and problems can only be fragmentarily interpreted by quoting his words.

Dale was reluctant to contribute to a book called *What I believe* compiled by Sir James Marchant [1953]. Dale's last sentence was:

'Sincerely, I cannot rid myself of the feeling that what I, in particular, believe, can hardly be of interest to anybody but myself.'

His attitude is perhaps best summarized by his reference to two quotations. The first is by Coleridge:

'He who begins by loving Christianity better than truth, will proceed by loving his own Sect or Church better than Christianity, and end in loving himself better than all.'

The second is by Leo Tolstoy, quoting the saying from Coleridge:

'I travelled the contrary way. I began by loving my Orthodox faith more than my peace, then I loved Christianity more than my Church, and now I love truth more than anything in the world.'

This resulted in the excommunication of Tolstoy from the orthodox church.

Reconciling the scientific with the religious state of mind, he expressed the 'creed' of the true scientist in the following words:

'The conflict started, according to my reading of the facts, not with an attack by science upon religion, but, on the contrary, with an attempt to invoke the authority of religions in order to restrict the freedom of natural science to pursue its quests for those truths which its own methods could discover, and within its own proper sphere of inquiry.'

Finally he concluded:

'For the purpose of science to allow belief to be established by tradition is dangerous and to accept a creed is suicidal. Religion by contrast, claims acceptance of a truth which, having been divinely revealed, must, in its essence, be regarded as fixed and final.'

His following words could have been written today in another context:

'… I find myself unable to believe that any knowledge which science is yet in a position to offer provides an adequate basis for that ethical sanction which bids us admire certain qualities of thought, character or conduct as virtuous and beautiful, and to condemn others as evil and ugly.'

It is also remarkable how he approached the problem of freedom of will:

'Not one of us, in the ordinary conduct of his life, can help acting, or thinking, as though he were free to choose and to decide. If this sense of free will be, then, indeed an illusion, it seems to be one from which none of us can in practice escape. Theology may feel called upon to find a way out of this dilemma; but science, content to wait for the advance of knowledge to remove, if it can, an apparent contradiction, will continue meanwhile its researches into material phenomena, within the logical frame of determinism which common experience seems to reject from the mind.'

He is sympathetic towards those who gladly embrace Christian ethics, but finds it difficult to accept Christian dogmas such as the existence of a Divine Plan, or the survival of the individual personality after body and brain have become dust:

'If a friend asks me to stand godfather at the christening of his infant son, is it proper that, unless I refuse to act, I should find myself, a man of science and a physiologist, committed to a public declaration that "I steadfastly believe" in "the resurrection of the flesh", to mention only one of the Articles of the ancient Creed?'

This statement might be regarded as offensive by some, but the following statement clearly defines his point of view:

'I am bound to believe for many like me today, the gentle, beneficent, heroic figure of the young Nazarene, and the simple nobility of His message, stand forth with a stronger appeal and in a clearer light when they can be freed from the clouds of supernatural legend and dogma, in which they were so soon and so naturally involved by the devotion of imaginative enthusiasts, in an age when all men were credulous, and eager for the miraculous, and some of them for the subtleties of theological speculation.'

He repeats this in his concluding words:

'If I were pressed for a statement of my personal conception of the God whose fatherly care, not only for mankind, but for all animated nature, Jesus proclaimed, I should feel tempted to take refuge in such a reply as St. Augustine made, to one who inquired concerning his conception of Time: "Until you asked me, I might have thought that I knew." Or more cowardly still, I could even claim orthodox warrant for replying: "Incomprehensible".'

Loewi's Reflections

In contrast to Dale, Loewi never produced an article dealing specifically with his religious views. His views, based on his Jewish confession, can at best be inferred from several sentences he wrote dealing with this subject, which can be found in two articles.

In a discussion about the nature of a Jew in a German environment, Otto Loewi said: 'I always considered myself as a German-speaking cosmopolitician of Jewish descent.' Concerning assimilation, he stated: 'All that matters is that the process of assimilation should have certain limits and that dignity and self-respect be preserved.' He was highly respected by his colleagues and coworkers, although he missed no opportunity to stress his consciousness of being a Jew [see Lembeck and Giere, 1968]. As member of the Jewish Community in Graz, Loewi was not personally active. Religious instruction of children in Austrian schools was, and still is, the responsibility of the clergy of each confession. Religious instruction was obligatory and was included in the annual school report. The majority of the children were Roman Catholic, and received religious instruction during the morning, whereas the non-Catholics, i.e. the few Protestants and Jews, were taught in the afternoon by the pastor or rabbi, respectively. Loewi's wedding ceremony did not take place in the Synagogue, but it was conducted by the rabbi in Loewi's home. The different confessions treated each other with mutual respect. The Loewi family even had a Christmas tree like all families in the neighbourhood – it would have been disappointing for the children not to have their own tree.

Concerning the essence of life, Loewi wrote:

'The essence of life, consisting in the tendency and potency of the organism to preserve it by all kinds of adjustments, cannot be explained by chemical or physical means. I cannot imagine that there exists any biologist nowadays, who would not approve of this statement. I am convinced that even during the epoch when the crude mechanistic concept of life predominated, most biologists must have been aware that teleology exists in living beings. How could it have been otherwise? The teleological view has at all times presented the starting point in the study of the chemical and physical factors involved in the single functions of the organism and in their coordination in order to relate them to their purpose. And yet almost all biologists up to this day have been reluctant to express or even touch on the teleological viewpoint… I am not a bit astonished that we are not able to understand this organisation of living matter. How could we expect such an understanding? We know that our perceptions are dependent on our sense organs. They therefore must be limited and imperfect. In addition, how could it be that our brain should be the object of understanding and at the same time the only tool for this understanding? Should we be disappointed about this limitation of human capacity? I do not feel this way. I am deeply thankful that there remains something great to be admired and to be revered, namely that inconceivable creative power of nature, which really gives rise to a deep religious feeling.'

In his lecture 'The Organism as a Unit', we find the following paragraph:

'What is life? St. Augustine when asked the meaning of time, said: "Si nemo ex me quaerat, scio; si quaerenti explicare velim, nescio." [When nobody asks me about it, I know it, but when I want to explain the things I have been asked, I don't know it.] We are faced with a similar situation if asked for a comprehensive definition of life. We may feel what life is, but we do not know how to express this feeling; hence we must content ourselves with examining marks which appear to characterize life in contrast to the non-animate.'

Dale also quoted these words of St. Augustine.

At the end of this lecture Loewi writes:

'Summarizing we have to make the statement that our knowledge concerning the higher central nervous functions which are so particularly important for the harmonious cooperation in the organism shows a great many gaps.

In spite of this I felt the urge to discuss the question to some length, may be for the reason that as one grows older one becomes much more impressed by the huge bulk of what we do *not* know, as compared with what we *do* know.'

Finally he writes:

'Hence, as far as the knowledge of last things is concerned, there does not exist a difference in principle between scientists as such and believers in individual religions. Both groups furthermore have a certain belief or faith: the scientist believes in the rationality of the universe as manifested in its ordered regularity. This aspect creates his admiration and awe. Goethe said: "It is the supreme satisfaction of the thinking man to explore the explorable and silently to revere the unknown." From reverence and awe arises a feeling which in my opinion can solely be defined as a religious feeling.

This tends to show that the scientist follows a drive not only of his mind but also of his spirit. Einstein, to whom it seems nature has given away more of its secrets than to anybody else, calls this cosmic religious experience the strongest and noblest mainspring of scientific research.'

Loewi used the word 'teleology' to describe the dogma of purpose and efficacy in nature which includes man, as originally formulated by Aristotle and which has continued to preoccupy philosophers throughout the ages, to discuss a variety of topics dealing with scientific and religious matters. In our time it has been subsumed under what Jacques Monod [1970] describes as *Le hazard et la nécessité*. Loewi was fully aware that our life includes many areas beyond the scope of scientific explanations. He fully acknowledged, and even admired, these aspects of life which also include religious matters. Dale's and Loewi's statements are certainly different, but their views are virtually identical.

References

Lembeck F, Giere W: Otto Loewi: Ein Lebensbild in Dokumenten. Berlin, Springer, 1968, p 48.
Marchant J: What I Believe. London, Odhams Press, 1953, pp 67–82.
Monod J: Le hazard et la nécessité. Paris, Seuil, 1970.

WILHELM FELDBERG

Wilhelm Feldberg at Mill Hill 1962.
Photo by K. Fleischhauer.

WILHELM FELDBERG – INTRODUCTORY REMARKS

In 1969, as president of the German Pharmacological Society, I (F.L.) had the honour to welcome Wilhelm Feldberg at the meeting in Mainz and to present the Schmiedeberg Plakette to him. Some years later, between 1972 and 1975, Sir Arnold Burgen was the President of IUPHAR (International Union of Pharmacology) and I was his Secretary-General. This resulted in frequent 'business meetings' at the National Institute for Medical Research at Mill Hill. After having completed our agenda, we always went for a short visit to Feldberg's laboratory. We often found him in his small office next to his laboratory, where usually an Indian scholar was sitting in front of a cat with cannulae implanted in its brain ventricles. When I used to come with Sir Arnold, he welcomed me with the words: 'Ah, Lembeck now you seem to be already a grown-up person!' – and we enjoyed this typical Feldberg joke.

Feldberg's working life lasted for 65 years and ended when he was 90 years old, when animal rights activists as 'advocates of animals' gained access to his laboratory, and took advantage of his open information about how he conducted animal experiments by drawing up a report for the Home Office. The alleged incidents led to his license being revoked. His long career came to an unhappy end. He died on October 23, 1993.

Bisset and Bliss [1997] in their memoirs on Feldberg review many details of his outstanding personality and his enormous scientific work, resulting in 350 publications. The use of the eserinized leech muscle, exquisitely sensitive to acetylcholine, made it possible to investigate the transmitter functions of acetylcholine in all parts of the nervous system. It resulted in 24 publications with Dale, Marthe Vogt, Gaddum, Burn and others within 3 years. These results and all his other lines of experimental research made him one of the greatest neuropharmacologists of the 20th century. Feldberg's excellence is complimentary and not secondary to that of Loewi and Dale.

The memoirs were supplemented by an addition by S.M. Hilton who was a student of Feldberg at Cambridge in the early 1950s and later a coworker at Mill Hill. He describes his outstanding gift as a teacher. He communicated his enthusiasm for his subject, his excitement even, to his delighted audience. Feldberg, during his time at Cambridge, once a week showed experiments as part of his lectures; these were unforgettable performances. Everyone, who has demonstrated pharmacological experiments to students, knows that such occasions remain firmly in the memory of the audience, unlike the transient memory which results from viewing them on a monitor.

His junior coworkers were put through a gruelling course, as he used to insist on them meticulously rehearsing all talks and communications, sometimes several times over. No theatrical producer could have been more exacting. Hilton remembered that his word-by-word examination of our manuscripts likewise was not less thorough.

A mirror of Feldberg's lectures might possibly be his reminiscences of an eyewitness in 'The Early History of Synaptic and Neuromuscular Transmission by Acetylcholine' which is included below from the booklet *The Pursuit of Nature* by Hodgkin et al. (see pp. 117–134); it seems to be available in only a few libraries.

How Wilhelm Feldberg Joined Sir Henry Dale

Although this is mentioned in the publication cited above, I would like to draw the reader's attention to the dramatic episode:

'One day in 1933, shortly after Hitler came to power, the Director of the Institute in Berlin where I was working sent for me and informed me that I had been dismissed, must leave the Institute at the latest by midnight that day, and was not allowed to enter it any more.

A few weeks later someone told me that the representative of the Rockefeller Foundation was staying in Berlin and that I should try to see him. I succeeded. He was most sympathetic, but said something like this: "You must understand, so many famous scientists have been dismissed whom we must help that it would not be fair to raise any hope of finding a position for a young person like you." Then, more to comfort me, "But at least let me take down your name. One never knows." And when I spelt out my name for him, he hesitated, and said, "I must have heard about you. Let me see." Turning back the pages of his diary, he suddenly said, delighted himself: "Here it is. I have a message for you from Sir Henry Dale whom I met in London about a fortnight ago. Sir Henry told me, if by chance I should meet Feldberg in Berlin, and if he has been dismissed, tell him I want him to come to London to work with me. So you are all right", he said warmly. "There is at least one person I needn't worry about any more".'

When Feldberg arrived in London, there was an actual need to measure acetylcholine released by nervous stimulation in other organs and species besides the frog heart. For doing this, he had brought a suitable present from Berlin, the eserinized leech muscle, exquisitely sensitive to acetylcholine. Führer [1917], who had discovered this, regarded it as suitable only in toxicological investigations. But Feldberg had already used it in Berlin to show the appearance of acetylcholine in the venous outflow from the adrenals after splanchnic nerve stimulation [Feldberg and Minz, 1933]. Feldberg and Krayer [1933] measured acetylcholine in the venous outflow from the heart of the dog after vagus nerve stimulation. The eserinized leech muscle made it possible to show the release of acetylcholine from autonomic ganglia and motor nerve endings. It resulted in 24 publications, covering the whole field within the following 3 years. It was even possible to deduce the amount released in response to a single nerve impulse at a single synapse in a sympathetic ganglion to as many as 3,000,000 molecules of acetylcholine [Feldberg and Vartiainen, 1934].

In their memoirs of Feldberg, Bisset and Bliss [1997] review his outstanding personality and his great scientific work; a contribution by S.M. Hilton has also been added since he is somebody who had the privilege to be in the audience during Feldberg's unforgettable lectures and demonstrations. Feldberg's work resulted in 350 publications, not counting his stimulating discussions with collaborators and visitors, his speeches full of witticisms, his jokes and anecdotes. Feldberg's excellence is complementary and not secondary to that of Dale and Loewi.

References

Bisset GW, Bliss TVP: Wilhelm Siegmund Feldberg, CBE. Biogr Mem Fellows R Soc Lond 1997;43:143–170.

Feldberg W, Krayer O: Das Auftreten eines acetylcholinähnlichen Stoffes im Herzvenenblut von Warmblütern bei Reizung der Nervi vagi. Arch Exp Pathol Pharmakol 1933;172:170–193.

Feldberg W, Minz B: Das Auftreten eines acetylcholinähnlichen Stoffes im Nebennierenvenenblut bei Reizung der Nervi splanchnici. Pflügers Arch Ges Physiol 1933;233: 657–682.

Feldberg W, Vartiainen A: Further observations on the physiology and pharmacology of a sympathetic ganglion. J Physiol 1934;83:103–128.

Fühner H: Ein Vorlesungsversuch zur Demonstration der erregbarkeitssteigernden Wirkung des Physostigmins. Arch Exp Pathol Pharmakol 1917;82:81–85.

THE EARLY HISTORY OF SYNAPTIC AND NEUROMUSCULAR TRANSMISSION BY ACETYLCHOLINE: REMINISCENCES OF AN EYE WITNESS

W. FELDBERG

Reprinted with permission from: The Pursuit of Nature, Hodgkin et al. eds.,
Cambridge University Press, 1977, pp. 65–83

Did it Really Begin in 1933?

It did, and yet it did not, because for the true beginning we must go back to Dale's early publication in 1914 on 'The action of certain esters and ethers of choline', in which he made the fundamental distinction between two types of action of acetylcholine, a muscarine and a nicotine action: that some of the peripheral actions of acetylcholine resembled those of muscarine, others, those of nicotine. Muscarine actions were those on smooth muscles, heart muscle and gland cells, and they were sensitive to atropine; nicotine actions were those on sympathetic (and parasympathetic) ganglia and on the cells of the adrenal medulla, and they in turn were abolished by large paralysing doses of nicotine. Nothing was said at that time about the third nicotine action of acetylcholine: that on motor endplates of striated muscles. This action was not discovered until much later.

Dale's early paper showing the distinction between muscarine and nicotine actions of acetylcholine is the true beginning of the story of acetylcholine in synaptic and neuromuscular transmission. The paper is important historically also because of two predictions or glimpses into the future which later were turned into facts by work carried out in Dale's own laboratory.

Dale predicted by sixteen years the presence of a cholinesterase in blood when he pointed out that the extraordinary evanescence of the effects of acetylcholine injected intravenously was probably due to its hydrolysis into choline and acetic acid by an esterase in blood. The enzyme was discovered independently by Matthes in Dale's laboratory, and by Engelhardt and Loewi in Graz. In both publications which appeared in 1930, eserine was used to inhibit the enzymic destruction of acetylcholine. Yet the obvious conclusion was not drawn that eserine might provide a simple means for detecting acetylcholine released into the blood.

It was a glimpse into the future, his statement in 1914 that acetylcholine would be the ideal transmitter substance for parasympathetic nerve effects, if there were any evidence for its presence in animal tissue. This evidence was provided by Dale himself together with Dudley fifteen years later when they identified acetylcholine in extracts of horse spleen.

But we find no suggestion, not even a hint that the nicotine actions of acetylcholine, too, might have physiological functions. Such a possibility was not envisaged in 1914, nor later when acetylcholine was shown to be a natural constituent of animal tissue, and when the presence of a cholinesterase in blood was discovered. The idea did not even arise when the contractures produced in denervated mammalian striated muscles by arterial injections of acetylcholine were investigated, for instance, in 1930 by Dale and Gaddum in the denervated tongue, and a year later by von Euler and Gaddum in the denervated muscles of the upper lip. The aim of these experiments was to shed light on the contractures produced in these denervated muscles on stimulation of parasympathetic and sympathetic nerves, and to find an explanation for this strange behaviour of nonmotor nerves in becoming 'pseudo-motor' as it was called, which had been known for so long. In 1863, the contracture produced in the denervated tongue on stimulation of the parasympathetic fibres in the lingual nerve was discovered by Philippeaux and Vulpian; it was confirmed by Heidenhain in 1883, and since then has often been referred to as the Vulpian-Heidenhain phenomenon. The contracture of the muscles of the upper lip on stimulation of the cervical sympathetic was described in 1885 by Rogowicz. The investigations of Dale, Gaddum and von Euler were a beautiful analysis and clarification of these pseudomotor phenomena, suggesting that they were due to acetylcholine released on stimulation of nerve fibres belonging to the autonomic nervous system, but the thought did not arise that acetylcholine might be the neuromuscular transmitter. They did not even try their method of arterial injections of acetylcholine on the non-denervated muscle. Had they done so and had they injected acetylcholine in larger doses they would have observed real contractions. Would that have been the moment for the idea to be born of a physiological role of the nicotine action of acetylcholine? Or would they have described the effect just as a third nicotine action of acetylcholine, emphasizing perhaps, as I did a year later when I observed these contractions, that denervation produced two changes in the response to acetylcholine, not only rendering the muscle more sensitive to it but also converting the response from contractions into a contracture.

How near we all were just before 1933 to the idea of a physiological role for the nicotine actions of acetylcholine, and yet how remote from this advance.

Today this may seem strange but, considering the ideas prevailing at that time about synaptic and neuromuscular transmission, it is not. We must remember that even the theory of chemical transmission from the endings of sympathetic and parasympathetic fibres to smooth muscles, heart muscle and gland cells was not yet universally accepted, although Loewi's experiments had gained wider and wider acceptance with each year. But the idea of chemical transmission occurring at the ganglionic synapse or at the neuromuscular junction was against the views of the electrophysiologists working in this field. They were convinced, and able to convince their colleagues, that with electrical methods alone it would be possible to unravel the mysteries of these transmissions, taking it for granted they could only be electrical, brought about by the eddy currents of the nerve impulse. Today we admire the ingenuity of their theories explaining these transmissions in electrical terms, the detonator response of Eccles, or the theory of Lapicque, that for transmission to occur the chronaxie of the nerve fibre must be the same as that of the innervated tissue.

Even after the first experimental evidence was obtained for the role of acetylcholine in these transmission processes, it took several years for the electrophysiologists reluctantly to give up their prejudice against a 'pharmacological interference' in these transmissions. Any result contradicting the acetylcholine theory was greeted with satisfaction. Lorente de No published experiments on the perfused superior cervical ganglion in which he found that acetylcholine appeared in the venous effluent on antidromic stimulation of the sympathetic fibres emerging from the ganglion and on stimulation of the vagus. If these results had been confirmed – but they were not (MacIntosh, 1938) – they would have provided decisive evidence against the role of acetylcholine in synaptic transmission. When Lorente de No's publication appeared, both Eccles and I were still in Australia. A colleague from Eccles's lab in Sydney came to see me in Melbourne. He said he had to give me, instead of greetings, the following message from Eccles: 'Acetylcholine is all wet!' I replied by telegram, 'Prefer wet acetylcholine to dry eddy currents'; signed, 'Anonymous'.

Many years later, at a meeting of the Physiological Society at University College London, after a communication about experiments which were meant to show that acetylcholine could not be the neuromuscular transmitter, G.L. Brown called from the Chair. 'Feldberg, you will have to fight with your back to the wall.' My reply was something like, 'I prefer this to any fighting without acetylcholine.'

When the acetylcholine theory became more firmly established, forgotten statements were dug up to show that there had been scientists in the past who had considered and favoured chemical transmission in these events. In 1937, Dale published a note in the *Proceedings* of the Society on 'Dubois-Reymond and chemical transmission' to illustrate that 60 years ago, in 1877, Dubois-Reymond had quite clearly suggested and favoured a chemical transmission of the excitation from motor nerve endings to striated muscle. Then there was the brilliant anticipation of this mode of action by T.R. Elliott, who had been the first to conceive the idea of chemical transmission in the autonomic nervous system. In his Sidney Ringer Memorial Lecture, he discussed this possibility also for striated muscles and the electric organ of fishes; he even made an attempt to extract the active substance. 'I have tried in vain to discover an active substance in the muscle plates of striped muscles. And Professor Herring was also disappointed when he examined for this purpose the electric organs of the skate which are exaggerated motor end plates ... But it is hard to forgo the belief that such discoveries lie in the lap of the future.' That was written in 1914. When, in 1942, Fessard and I wrote our paper 'On the cholinergic nature of the nerves to the electric organ of the Torpedo', Dale drew my attention to Elliott's lecture and we quoted his prediction. I was so impressed by it that I was anxious to meet Elliott, and Dale, who knew him well, invited us both for dinner one evening at the Athenaeum. There I asked Elliott what kind of extraction they had used. As far as he could remember they had simply used water or saline extracts. If only they had used hydrochloric acid and thus prevented the spontaneous and enzymic hydrolysis of acetylcholine they could hardly have missed the high activity in the extracts from the electric organ. What a fantastic near-miss it had been!

How Did it Begin in 1933?

And how did it happen that I was in Dale's lab at that time? This, at least, I have no difficulty in answering.

One day in 1933, shortly after Hitler came to power, the Director of the Institute in Berlin where I was working sent for me and informed me that I had been dismissed, must leave the Institute at the latest by midnight that day, and was not allowed to enter it any more.

A few weeks later, someone told me that the representative of the Rockefeller Foundation was staying in Berlin and that I should try to see him. I succeeded. He was most sympathetic, but said something like this: 'You must understand, Feldberg, so many famous scientists have been dismissed whom we must help that it would not be fair to raise any hope of finding a position for a young person like you.' Then, more to comfort me, 'But at least let me take down your name. One never knows.' And when I spelt out my name for him, he hesitated, and said, 'I must have heard about you. Let me see.' Turning back the pages of his diary, he suddenly said, delighted himself: 'Here it is. I have a message for you from Sir Henry Dale whom I met in London about a fortnight ago. Sir Henry told me, if by chance I should meet Feldberg in Berlin, and if he has been dismissed, tell him I want him to come to London to work with me. So you are all right,' he said warmly. 'There is at least one person I needn't worry about any more.'

Here I must explain that I had known Dale since 1927, when I was a visiting scientist in his department. In 1930, Schild and I had dedicated a book to him on histamine, and in 1932, I met Dale again in Wiesbaden at a meeting of the German Pharmacological Society followed by the German Internisten Kongress where Dale was the invited lecturer. At the meeting of the Pharmacological Society (Otto Loewi was there, too) I gave a communication on the release of an acetylcholine-like substance from the tongue into the blood during stimulation of the chordalingual nerve, illustrating how easy it had become to detect the released acetylcholine once its destruction was prevented by an intravenous injection of eserine and the venous blood was allowed to pass over the eserinized leech muscle preparation. Another communication on the leech muscle preparation itself was given by Minz who at that time was working with me in the Physiological Institute of the University of Berlin. Dale was greatly interested in our communications.

Having received Dale's message from the representative of the Rockefeller Foundation, I wrote at once to Sir Henry, immediately received an invitation, and as soon as all necessary formalities were completed, I was literally thrown out by my wife who, with our two small children, stayed behind to get everything ready and packed for what might, and eventually did, become an emigration. I landed at Harwich on 7 July 1933.

This is how I happened to be with Dale in 1933, and worked in is laboratory for over two and a half years until we left for Australia which, as shown by my Certificate of Registration, was on 15 April 1936.

The first weeks in London were terrible, with the constant fear something might happen to the family in Germany and prevent them from leaving. When they finally arrived, I traveled to Harwich the night before to meet the boat due to land at about five the next morning. One could not help being in a state of anxiety because it was believed that passengers were not infrequently taken off the train at the Dutch frontier and sent back. Hours before the boat was due I walked up and down the empty quay, looking into the empty rooms until they filled with customs officials. One of the immigration officers must have observed me, because when he handed the landing permit to my wife, he asked her if it was me who had been waiting outside, and then said, 'Mrs Feldberg, you must never again leave your husband alone.' That was the compassion shown at that time by a large part of the English people to the refugees from Germany: that's what we had become.

But the following two and a half years were an unbelievably happy time. Naturally one was deeply concerned with the fate of relations, friends, colleagues, and all those who were still in danger, and one tried to help whenever possible. But one could not help being happy; it was like a new life after a severe illness: to be safe with the family and in addition to have the possibility to do research under the most favourable conditions imaginable. Later, colleagues have often said to me that those years in Hampstead after 1933 must have been an exciting time. I, personally, have never been able to separate from each other our happiness and the work going on in the lab.

We had rented a two-roomed furnished flat about ten minutes' walk from the Institute in Hampstead, and nearly every morning my wife accompanied me to the lab, wished me good luck for the experiment and asked me not to worry however late it might be in the evening, just to ring her before leaving and she would meet me half way. One morning on the way to the lab we had stopped in front of a fishmonger's shop when Lady Dale passed, and asked why we were staring at lobsters. My wife explained that I was so fond of them, that whenever I felt we had done a really exciting experiment resulting in a definite advance, we would celebrate it with a lobster. When Sir Henry came into the lab next morning, the first thing he said was that, from then on, lobster experiments would be celebrated in his house. Sometimes when Sir Henry had to be away during the day at some meeting and could not be with the experiment, he would rush into the lab in the evening before dinner and ask, 'A lobster experiment today?' Synaptic and neuromuscular transmission have been responsible for a good number of lobster dinners!

Dale liked to start early in the morning, and realizing that I would most probably be late, I was often commanded to have breakfast with him at Mount Vernon House, to make certain the experiment would begin on time.

For the first six months I could support myself, then my accounts in Germany were frozen, but Dale obtained a Rockefeller grant for me for a year, which was paid quarterly. One morning, near the end of a month, Dale came into the lab whilst I was preparing our cat, and asked how often I still expected a quarterly payment. I said, 'once more'. Five minutes later, he came back and said, 'you were wrong, you have had your last payment; how long will you be able to go on?' I said, 'for another fortnight'. Collison, Dale's technician, who was assisting me, got really worried and asked what I was going to do. 'Let's go on with the experiment,' I said, 'I can't do anything; either Sir Henry succeeds in finding financial support, then all is well. Otherwise it is even more important that we do a few more good experiments in the time that is left.' Those were happy days. We had learned not to worry about trivial things. Dale naturally succeeded in getting the Rockefeller grant prolonged for another year, but he was furious with me that I had not been able nor willing to take the matter seriously.

Lady Dale told my wife how glad she was that we had come to England, because before then her husband had stopped working in the lab and now he was enjoying doing experiments again. The best proof that Dale had enjoyed the time I worked in his lab is the fact that before he retired in 1940, he told me that if it had not been for the war, he had hoped, and would have liked best, to work with me again after his retirement. Dale's last experimental paper before 1933 was with Bauer, Poulsson and Richards, and on an entirely different subject, on liver perfusion. It appeared in the April number of the *Journal of Physiology* in 1932 and must therefore have been sent in some time in 1931. The one prior to this was the one with Gaddum in 1930 on acetylcholine contractions in denervated mammalian striated muscles. After I left for Australia, Dale again stopped doing experiments.

Between 1933 and 1936, the index volume of the *Journal of Physiology* for volumes 61 to 100 lists, from Dale's laboratory, fourteen publications, printed communications or full papers, on acetylcholine in ganglionic and neuromuscular transmission, and another ten publications on transmission by acetylcholine in other tissues: in the suprarenal medulla, stomach wall, sweat glands, salivary glands and central nervous system. I cannot help it, but my name appears as co-author on all twenty-four of them. What is the explanation? And what was my contribution which resulted in this embarrassing fact, that I was an eye-witness in all of them?

To make use of a metaphor: perhaps it was that I had brought with me a key that would open the doors. Dale and Gaddum seemed to know what lay behind them, but I had the key. So I was asked to open first this one, then that one and so on, one after the other; and I never refused to do so. And what was the 'key' that I had brought with me? It was my method of detecting acetylcholine released during nerve stimulation. It was the use of eserine, its intravenous injection or, in perfusion experiments, its addition to the perfusion fluid in order to inhibit the enzymic destruction of the released acetylcholine and then to detect it in the venous effluent, blood or perfusate, by means of a simple, specific and sensitive test, the eserinized leech muscle preparation. In the historical abstract to my demonstration on acetylcholine release from the perfused sympathetic ganglion, I have told the story of how it came about that this preparation became our routine test for assaying minute amounts of acetylcholine. With this 'key' we obtained the first direct experimental evidence for the role of acetylcholine both in ganglionic and neuromuscular transmission. The second evidence followed in quick succession. It, too, was based on the effect of eserine to inhibit the enzymic destruction of acetylcholine. Eserine should affect the response to nerve stimulation, if due to acetylcholine, in a way consistent with persistence of undestroyed acetylcholine at the site of release. Eserine had been shown before to affect the response to nerve stimulation in this way but it had not been used so persistently for this purpose as we now used it, in order to obtain evidence for the role of acetylcholine in transmission processes. During these experiments it became evident that the effect need not necessarily be a potentiation of the response; eserine could have the opposite effect and produce attenuation due to a paralysing action of excess acetylcholine, for instance in the sympathetic ganglion.

With this two-fold experimental evidence the cholinergic nature of ganglionic and neuromuscular transmission was actually established although it took years before it was generally accepted. For the conversion of the strongest opponent, Eccles, we had to wait for over ten years. Those interested in the controversy during these intervening years cannot do better than to read the charming little book by Bacq on *Chemical Transmission of Nerve Impulse* which was presented to the participants of the Sir Henry Dale Centennial Symposium in Cambridge on 17–19 September 1975. Bacq dedicates a chapter to the controversy. He remarks that in 1943 both Eccles and I had a paper in the *Journal of Physiology* in which we discussed the problem of synaptic transmission. Eccles does not mention a single time the term acetylcholine, but states categorically that it has now been established that neuromuscular and synaptic transmission are mediated by the local negative potential set up by the nerve impulse. I concluded from experiments on acetylcholine synthesis that such synthesis is a property of the preganglionic endings in sympathetic ganglia and necessary for normal and particularly for sustained synaptic transmission. Bacq writes, 'In his work Feldberg ignored Eccles' arguments, and he did not even bother to refute them. The two groups knew one another, but there was a wall between them; each pursued his own research with his own basic concepts and methods. The ability to consciously exclude each other's views so rigidly was astounding, the more so in that the two papers appeared in the same issue of the *Journal of Physiology*.' Bacq was right. But why should I read Eccles? He did not believe in acetylcholine and we were convinced of its role in synaptic and neuromuscular transmission. As Taoism tells us, a good man does not argue!

The strong opposition of Eccles and others, however, exerted a most beneficial effect. We were not allowed to relax, but were forced to accumulate more and more detailed evidence in support of our theory.

Perhaps I might illustrate by another instance, the difficulties electrophysiologists had at that time, in switching over completely in their way of thinking from electrical to chemical transmission, even when accepting the transmitter role of acetylcholine. I referred to experiments on the electric organ of the *Torpedo* by Fessard from Paris, and myself: In 1939, he invited me to work with him at the maritime station in Archachon, in the South of France, so that we might apply to the electric organ the methods which had proved so successful in demonstrating the role of acetylcholine in neuromuscular transmission.

The electric organ of the *Torpedo*, as well as of other fish, may be regarded as a collection of modified and exaggerated motor endplates lacking the contractile structure of voluntary muscle fibres. In the *Torpedo*, the organ consists of a great number of prisms arranged side by side, each prism being built up of several hundred superimposed plates all of which are covered on their ventral side by a terminal nerve net. At the moment of the discharge the ventral sides of all plates become negative to the dorsal nerve-free sides; the discharge produced by a whole organ may amount to several hundred volts. Auger & Fessard (1938, 1939) had done some fundamental work on the isolated prisms and, from the latency and form of the discharge on electrical stimulation, had been able to conclude that it is produced indirectly and that the organ cannot be activated electrically by the impulses reaching the terminal network. In addition, there was the very suggestive fact that the organ was an extremely rich source of cholinesterase (Marnay, 1937; Nachmansohn & Lederer, 1939).

Fessard and I having found that the organ was also rich in acetylcholine and that on stimulation of its nerves acetylcholine appeared in the effluent when the organ was perfused through its artery with the appropriate salt solution, which naturally had to contain eserine, we also wanted to know if acetylcholine had the electrogenic property required to qualify as transmitter. So the perfused organ and I were imprisoned in a small screened cubicle whilst the recording was done in freedom outside. When all was ready for the recording, I was given the word 'go' and at once made my arterial injection of acetylcholine. It resulted in an angry shout from outside. I was told I had nearly wrecked their apparatus. I had to be insulated, had to put on rubber gloves, and when that did not help, rubber boots as well. But even that did not help, and the engineer in charge of the recording refused to go on, being afraid for his equipment. I begged him to allow me just one more injection, and when it was made it was greeted with a joyful shout. The engineer opened my prison and said beamingly: a perfect injection, and absolutely no effect. I had injected the salt solution without acetylcholine. So I told him what had been wrecking his apparatus was my acetylcholine. From the discussion which followed, in rapid French, unintelligible to me, between Fessard and him, it came out that the engineer did not believe in acetylcholine, but he had wanted to give us a fair chance and therefore had recorded with the highest sensitivity. When he reduced the amplification and allowed a fraction only of the discharge to enter the amplifier, and I reduced the dose of acetylcholine, its injection produced a beautiful deflexion, still too large to be recorded in its entirety, but no longer frightening to the engineer.

The real reason for referring again to our experiments on the electric organ, however, was a different one. When writing up the experiments, I suggested as title for our paper something like 'The cholinergic nature of the nerves to the electric organ'. But Fessard said, 'We cannot do this'. And when I asked why, he said, 'Because l'organ électrique n'existe pas'. I was dumbfounded and amused. Here we had been working for weeks on an organ which certainly weighed more than a pound, and then to be told it does not exist! But Fessard explained. 'You must realize,' he said, 'that the organ, once its nerves have degenerated, is inexcitable.' This was shown by Garten in 1910, and confirmed by Fessard himself when working on the isolated prisms. He showed that they discharge normally at the slightest provocation not only on electrical stimulation but also when lightly pressed upon, say, by the tip of a pencil. However, once the nerves to the organ have degenerated, it is impossible to obtain a discharge. The electric organ is thus but part of the nervous mechanism and does not exist on its own. This interpretation illustrates how easy it was at that time to fall back into thinking in terms of electrical transmission, because that is what it was. A relapse, but a momentary one, because it did not take us long to find the interpretation in terms of chemical transmission, that the sole physiological stimulus able to excite the electric organ and cause it to discharge is acetylcholine. Electrical and mechanical stimulation of the organ or its prisms does nothing but excite the nerve fibres or nerve nets at the ventral surface of the plates to release acetylcholine which then produces the discharge; once the nerves have degenerated there is no acetylcholine available to be released by whatever stimuli; therefore no discharge is possible. According to this interpretation arterial injections into the perfused organ should produce a discharge also after degeneration of its nerves. This we intended to show. Fessard was going to operate on some fish and a few months later when the nerves had degenerated, bring the survivors by tank to Paris. I, in the meantime, would try to obtain a grant enabling me to fly to Paris about once a month so that together we could perfuse the denervated organ. I made a bet with Fessard that acetylcholine would cause a discharge. A few days after my return to London, I saw Dale at a meeting of the Physiological Society, talking to the representative of the Rockefeller Foundation. When Dale saw me he waved me over, and I had to tell both of them our results. What a chance! The outcome was a grant to enable me to fly over to Paris once a month. I remember Dale asking me if I would do so also if one fish only survived. I could not resist replying that I would do so even if none survived! The opportunity to make use of the grant, however, never arose. War started, and although I am certain that Fessard and I had the same opinion about the outcome of our bet, I have not been able to collect it – my bottle, of champagne!

The Years 1933 to 1936

A few dates may illustrate the rapidity with which everything happened in 1933 and during the following months. The first publications from this period dealing with chemical transmission by acetylcholine were printed communications given to the meeting of the Physiological Society on 18 November 1933. As they had naturally to be sent in some time earlier, it meant I had not been working in Hampstead for much longer than three months.

One of the communications sent in was by Dale and myself on the vagal innervation to the stomach. The motor effects to the stomach, as those to the urinary bladder, were known to be resistant to the action of atropine. This had been an obstacle to the general acceptance of acetylcholine, or any choline ester, as the probable chemical transmitter of these parasympathetic effects. Dale's keenness to find out if such effects were also associated with release of acetylcholine was what started the experiments. In our communication we reported a manifold increase in the acetylcholine content of the venous blood collected from the stomach during vagal stimulation in dogs, provided naturally that they were treated with an intravenous injection of eserine.

Another communication was by Minz, Tsudzimura and myself on the role of acetylcholine as transmitter to the suprarenal medulla. Before I left Germany, Minz and I had found, with the help of intravenous eserine and of the eserinized leech muscle, that an acetylcholine-like substance appeared during stimulation of the splanchnic nerves in the venous blood from the suprarenals. We had continued this work with a Japanese colleague, Tsudzimura, when it was interrupted by my dismissal. But I was allowed in Dale's laboratory to continue and complete the work which provided the second piece of evidence for this transmission by acetylcholine: the potentiating effect of eserine on the pressor response produced by splanchnic stimulation to the suprarenals.

The third communication, by Gaddum and myself, was the one on synaptic transmission by acetylcholine in sympathetic ganglia. In my historical abstract to the demonstration of these experiments, I told the story of how it came about that we looked for acetylcholine in this transmission, but I shall tell it again. There were first two findings we had made independently of each other. The finding just mentioned, by Minz and myself, of the appearance of an acetylcholine-like substance in the blood from the suprarenals on splanchnic stimulation, that is, on stimulation of nerve fibres which correspond to preganglionic sympathetic neurons, and the pharmacological identification by Chang and Gaddum (1933) of acetylcholine in extracts of the sympathetic chain of the horse. They were much nearer the truth than we were when they suggested at that time that acetylcholine 'might play a part in normal transmission of impulses through ganglia'. How timid and vague this statement appears today! Then there was the additional strong impetus given by a publication of Kibjakow (1933) which had just appeared in *Pflüger's Archiv*, the description of a method for perfusing the superior cervical ganglion in a cat. Just what was needed! It is not so important today that his results have never been confirmed, that the effluent collected during stimulation of the cervical sympathetic when added to the fluid perfusing the same or another ganglion stimulated it. What was important and revolutionary at that time was his idea, and his attempt to test it, that a substance should be released from the preganglionic nerve endings and be responsible for the transmission of the nerve impulse across the synapse. Gaddum was fascinated and excited; he showed me Kibjakow's paper; I had to read it at once. But having done so, he had no difficulty in persuading me to join forces to use Kibjakow's method, but with the additional trick of adding eserine to the perfusion fluid, to test the effluent collected during stimulation of the cervical sympathetic on the eserinized leech muscle and, if it should cause contraction, to use the procedure Gaddum had so beautifully worked out with Chang, of identifying pharmacologically an active substance with acetylcholine by making parallel estimates on different assay preparations. And that was what we did.

But there was a fourth communication given by Dale at this meeting on 18 November 1933. It was the most important of all, partly the outcome of the findings reported in the other communications. It was his 'Nomenclature of fibres in the autonomic nervous system and their effects'. His introduction of the terms 'cholinergic' and 'adrenergic' fibres or neurones; the realization that the anatomical designation of sympathetic and parasympathetic fibres was not sufficient because it did not give any information of whether a nerve fibre acted through the release of acetylcholine or of 'something like adrenaline' which we now know to be noradrenaline. I remember well Dale having gone on several days to the Library, which was not usual for him, and then his happiness one day when he came to show me the typescript of his communication. He was fully aware of its importance.

With this nomenclature a new dimension was added to our understanding of the autonomic nervous system. It told us that probably all its preganglionic fibres, the sympathetic and the parasympathetic ones, are cholinergic and that there are exeptions to the general rule according to which the postganglionic fibres in the parasympathetic system are all cholinergic and those in the sympathetic system all adrenergic. One such exception mentioned in his communication was the innervation of the sweat glands, which in cats are found in the hairless pads of the feet. The secretory fibres to these glands are anatomically sympathetic, but as the sweating is abolished by atropine it appeared to be transmitted by something like acetylcholine. No wonder then that within weeks Dale and I perfused the cat's hind foot with Locke's solution containing eserine. And it came as no surprise to us that the sweating produced in the hairless pad on stimulation of the lower end of the abdominal sympathetic chain was associated with the appearance in the venous effluent of a substance which contracted the eserinized leech muscle and which had the other usual properties by which we recognized acetylcholine.

The nomenclature was given solely for fibres of the autonomic nervous system. No mention was made of the cholinergic nature of motor nerves to striated muscles. Yet again within months, Dale and I began perfusing with eserinized Locke's solution the cat's tongue, and found that stimulation of its motor nerve, the hypoglossus, with its sympathetic fibres degenerated, led to the appearance in the venous effluent of a substance which contracted the eserinized leech muscle and had the other properties of acetylcholine. We then obtained the same result on perfusion of the leg muscles of dogs on stimulation of the appropriate spinal ventral roots with the sympathetic fibres degenerated. These results provided the first experimental evidence for the cholinergic nature of neuromuscular transmission.

Certainly at the time when Dale wrote his printed communication none of us thought of this extension of cholinergic nerves to striated muscles. So when did the idea arise? I always thought it was some time after the meeting of the Society.

But in *Adventures in Physiology* Dale comments on the communication and tells us that at the meeting he 'made a preliminary, verbal mention of our plan to test the significance of these findings, by analogy, for the transmission from motor nerve endings to voluntary muscle fibres'. With 'these findings' he referred to our experiments on the role of acetylcholine as transmitter from preganglionic sympathetic fibres to the ganglion cells and the medullary cells of the adrenals which he described as a 'rather daring excursion'. Yet for those not inhibited by the thinking of electrophysiologists at that time, it was nothing special to look for a physiological function of two pharmacological actions which had been known for nearly twenty years. On the other hand, it is difficult today to realize the immensity of the novelty of looking, as Dale expressed it 'by analogy' for the role of acetylcholine in neuromuscular transmission, because the mammalian muscles were usually regarded as being completely insensitive to the action of acetylcholine when their motor nerve supply was intact. The recognized pharmacological effect was the contracture after motor nerve degeneration.

I mentioned the rapidity in the development of the acetylcholine theory in 1933 and during the following months and that five communications from Dale's laboratory were given at the meeting of the Physiological Society in November 1933. Six months later, another five were given at the May meeting in 1934, also dealing with the transmitter function of acetylcholine. Two are of no interest to our problem. The third and fourth dealt with the experiments just mentioned on the cholinergic nature of the sympathetic innervation to the sweat glands of the cat and of the motor innervation to voluntary skeletal muscle.

The fifth communication dealt with the 'Action of eserine in transmission through the superior cervical ganglion'. It gave the second piece of evidence required for the acceptance of the role of acetylcholine in ganglionic transmission. It showed that the contractions of the nictitating membrane produced by preganglionic stimulation were potentiated when eserine was added to the fluid perfusing the ganglion in high dilution, but depressed when added in stronger concentration; and that eserine had the same effects on the contractions produced by small doses of acetylcholine injected into the perfusion. This communication was by Vartiainen, a visiting scientist from Finland, and myself. We could also give the first estimate of the number of molecules of acetylcholine released by a single preganglionic volley per synapse. Calculated from the amount of acetylcholine appearing in the venous effluent of the perfused ganglion, the estimate was about three million molecules.

All these printed communications came out shortly afterwards as full papers in the *Journal of Physiology* except the one on neuromuscular transmission, which took longer. The reason was that we felt we had to amplify our results to avoid unnecessary criticism because we knew the opposition to our theory would be strong. In these experiments we were fortunate in being joined by Marthe Vogt who had left Germany because she detested the Nazi regime and what it stood for. During the months after her arrival we extended the limited results so far obtained to other muscles, including the leg muscles of the frog and, in order to meet possible criticism that the muscular contractions were responsible for the release of acetylcholine, we showed that the release did not occur on direct stimulation of the muscles, provided their nerves had degenerated, but that it occurred after curarine which abolished the contractions. Curarine was thus shown to act on motor nerve stimulation like atropine on parasympathetic stimulation, preventing not the release of acetylcholine but its action. All these results were included in the full paper by Dale, Vogt and myself which was sent to the *Journal of Physiology* in January 1936. This paper, however, did not yet contain experiments on contractions produced by acetylcholine or on the effect of eserine on motor nerve stimulation, but they were under way, and published four months later.

Gaddum had left at the end of 1933 to take up an appointment as professor of pharmacology at the University of Cairo. I still find it difficult to realize how short this fruitful time was, less than half a year, that we had worked together. Dale must have felt that it would become essential at some stage in the not too distant future to apply electrophysiological methods to the theory of ganglionic and neuromuscular transmission by acetylcholine. So he offered the position which had become vacant to G. L. Brown. He accepted and started working in Hampstead in April 1934.

In the *Biographical Memoirs* of The Royal Society, MacIntosh and Paton say that Brown always liked using electrical apparatus and that his experience of working with Eccles's amplifiers and oscillograph confirmed his preference. He had worked with Eccles in Oxford for half a year, but before he came to Hampstead he was working with McSwiney in Leeds on the innervation of the stomach. 'I was so interested in how the impulse got there and what happened to the inhibited or excited muscle that the process of neuroeffector transmission left me cold,' wrote Brown long afterwards. What better choice could have been made!

With Brown's arrival electrophysiological methods were introduced into Dale's laboratory. During his first month there, Brown was preoccupied with building, almost single-handed, his own amplifier for recording action potentials with the aid of a Matthews oscillograph. Later he divided his time between improving his equipment, which to my recollection seemed to go on indefinitely, and doing experiments with me on the superior cervical sympathetic ganglion, mainly perfusion experiments, and later also on neuromuscular transmission. He preferred electrophysiological methods for analysis but was always keen to try others as well. The main results we obtained together during the following months on ganglionic transmission, I think, were the following six, which are given in chronological order:

(1) Potassium, as well as rubidium, released acetylcholine from the preganglionic endings of the perfused ganglion.

(2) The ganglion cells were not only stimulated by potassium (this had been shown previously by Vartiainen and myself) but also, with larger doses, paralysed. The paralysing action on the cells did not affect the release of acetylcholine from the preganglionic endings.

(3) The ganglion became nearly depleted of its acetylcholine after degeneration of its preganglionic fibres.

(4) The action of curarine on the ganglion. It rendered the ganglion cells insensitive to acetylcholine, but not to potassium, and it did not prevent the release of acetylcholine from the preganglionic endings.

(5) Partial paralysis of the ganglion by excess released acetylcholine. In the presence of eserine the concentration of acetylcholine released in the perfused ganglion during the initial stages of stimulation was sufficiently high to do so.

(6) The release of acetylcholine during preganglionic stimulation was associated with synthesis of acetylcholine. During prolonged stimulation the amounts released were several times those obtainable from an unstimulated ganglion, yet its acetylcholine content did not diminish.

Another finding, though not a regular one, and the first of its kind concerning acetylcholine synthesis, was the ability of choline to augment the release of acetylcholine from the perfused ganglion when its output had gone down during prolonged stimulation.

Electrophysiological methods were not required to obtain these results, although some recordings of action potentials of ganglia were made and published, and the methods were used to study the effect of potassium injections on facilitation between two successive volleys, but nothing new was added to the problem of the role of acetylcholine in ganglionic transmission.

The superiority of the electrophysiological methods and the necessity to use them, however, became evident for the experiments on neuromuscular transmission. Before describing them, I shall mention in passing that one of the first indications for the role of acetylcholine in central transmission dates back to this period: the finding by Schriever and myself that after eserine injections, adrenaline and asphyxia caused the appearance of acetylcholine or its increase, in the cisternal cerebrospinal fluid of dogs.

The story of neuromuscular transmission by acetylcholine is an instance where a special physiological function of a substance was discovered before its corresponding pharmacological effect. Before 1934, acetylcholine contractions of normally innervated voluntary mammalian muscles were practically unknown. True, there were a few isolated observations of this kind, but little notice was taken of them and their significance was not recognized. However, once acetylcholine was found to appear in the effluent from a muscle perfused with eserinized salt solution during stimulation of its motor nerve, it became essential to 'discover' this action of acetylcholine, because how could acetylcholine play a role in neuromuscular transmission if it were unable to contract voluntary muscles. So we had to make a real effort to obtain this evidence.

Reading through the publication by Brown, Dale and myself, the memory comes back of the struggle we had, and how long it took us to obtain 'decent' contractions. In the course of our experiments we found that in muscles perfused with physiological salt solutions the conditions were never optimal because even in the earliest stages of perfusion an incipient oedema would dilute the injected acetylcholine solution. We realized we had to study not only the effect of acetylcholine under natural circulation of the muscle, but also to reduce the volume in which the acetylcholine had to be injected and to make the injections into the muscle artery as close to the muscle and as rapid as possible. The method of 'close arterial injection' as it came to be known was worked out to perfection later on by Brown when he used it for his classical electrophysiological analysis of the acetylcholine effect in different muscles. But we obtained the evidence needed, because we could show that acetylcholine produces muscular contractions.

In the beginning we recorded the contractions of the perfused gastrocnemius muscle of the cat mechanically with a tension lever on a smoked drum. I remember how excited we became when, in one experiment on close arterial injection of a relatively small dose of acetylcholine (20 μg) a rapid contraction developed with a tension of nearly 7 kg which was more than three times that produced by a maximal motor nerve twitch. Although the duration of the contraction was longer than that of a single maximal twitch it was certainly too short to be a contracture. The question which haunted us at the beginning was, could the response be a contracture? Today this may seem strange but at that time it was not, because only acetylcholine contractures were known. Therefore our excitement when, during experiments with natural circulation of the muscle, even smaller doses of acetylcholine produced contractions that on the smoked drum appeared to be as rapid, or nearly as rapid, as single maximal motor nerve twitches, and when on changing from the smoked drum to optical isometric myograms the contractions really proved to be of only slightly longer duration.

We could give the correct interpretation of our results, that the response consisted of a repetitive firing of the muscle fibres due to persistence of undestroyed acetylcholine at the motor end plates, and that any contraction produced by the acetylcholine injections whether it was stronger or weaker than a single twitch 'would have the nature rather, of a short, asynchronous tetanus'. In the proofs we were able to add that electrical records obtained in the meantime by G.L. Brown proved that our deduction was correct.

This interpretation had become obvious after we had analysed the potentiating effect of eserine on motor nerve twitches. How well I remember the very first experiment, and the next one as well, on the effect of eserine in a spinal cat. My suggestion was to use maximal single motor nerve stimuli. Brown was against it. 'What do you expect?' he said, 'the twitch is already maximal. You should know there is an all-or-none law!' 'But,' I said, 'what does it matter; let's see what happens; in all previous experiments on the release of acetylcholine we used maximal stimuli.' 'All right!' he said. And then our joyful surprise when, after the eserine injection, the response recorded on the smoked drum began to creep up and increased with each impulse to reach, during the following two minutes, double its original size. When I dared to remark that perhaps the all-or-none law did not apply, Brown became angry and retorted, 'Don't be a fool! Wait!' He had at once grasped the meaning of the result and was angry at not having tought of it before. In the next experiment we recorded the action potentials and saw: eserine converted the single 'twitch' into a waning 'tetanus'. 'What do you think now, Feldberg?' asked Brown triumphantly. 'You cannot remain a fool for long when working with great people.' What else could I reply? Brown and I loved to tease and to be rude to each other without personal offence. But we were all happy, having obtained so beautifully this last pieces of evidence which was necessary for the acceptance of the acetylcholine theory of neuromuscular transmission.

Epilogue

Events seen by two eye-witnesses look different. Everyone knows this. Further, no judge would accept, without the greatest reservation, the account of an eye-witness after forty years. Memory of recent events diminishes with age, but that is not so with events that have taken place many, many years ago. They may even become more vivid, but only some of them remain with us 'like pockets of snow in a general thaw'. In addition, memory often plays strange tricks. So my account of the events at Hampstead in the 1930s must be taken with all the reservations normally applicable to an eye-witness's account.

In his little book on *Chemical Transmission of Nerve Impulses*, Bacq raised the question of whether, without me, research on chemical transmission of nerve impulses might have taken a different course. In my view, the role of acetylcholine in ganglionic and neuromuscular transmission was due to be discovered in one laboratory or another. Most unlikely that I alone would have stumbled on it, except if, through results obtained in other laboratories my eyes would have been opened. Gaddum was nearer to the truth than I was. But I doubt that he would have started the ganglion perfusion experiments alone before leaving for Cairo, or later on under the less favourable conditions there for doing experiments, and with the new teaching responsibilities. In Cairo his main research interest shifted to histamine. Two printed communications, not followed by full papers, were all that appeared from him and his Egyptian colleagues on the role of acetylcholine in chemical transmission, and these publications did not add much to our problem.

On the other hand, it is most likely, in my view, that work in Dale's laboratory after 1933 would have brought to light the role of acetylcholine in the two transmission processes even without my presence. Dale knew of Gaddum's finding of acetylcholine in extracts of the sympathetic chain, of Kibjakow's ganglion perfusion, of my finding of the appearance of acetylcholine in the venous blood from the suprarenals during splanchnic stimulation, and of my successful use of eserine for the detection of acetylcholine released during nerve stimulation. If the possibility of a physiological role of the nicotine actions of acetylcholine lay dormant in the mind of any one person, that person would have been Dale. I could well imagine him asking the next visiting scientist to his lab to use Kibjakow's method of ganglion perfusion and later to perfuse a skeletal muscle with eserinized salt solution, thus getting the results we had obtained. On the other hand, one cannot be certain, because Dale had stopped doing experiments before 1933, and again after 1936. Dale loved to dwell on the role played by lucky accidents in his research. Perhaps my dismissal in Germany on account of Hitler was yet another of Dale's lucky accidents, certainly lucky for me.

People have often asked me if I was sad to leave Hampstead in 1936. Nearly every visiting scientist working in Dale's lab was sad to leave. In this sense I was sad, too, but not in any other way. I knew that for the further advance in our insight of the role of acetylcholine in chemical transmission, entirely different methods would be needed from those simple ones at my disposal, which had been so successful at the beginning of our story. I would no longer be able to contribute much, and I might have outstayed my usefulness. We left with the rare sense of happiness which comes through the knowledge of having been useful – in an exciting period of rapid scientific advance.

SUBSTANCES

From the History of Scientists to That of Neurotransmitter Substances

The reader may have found that I have concentrated on three key persons, on Loewi, Dale and Feldberg.

There are more biographic articles about Loewi than about the other two scientists. My personal relationship to Loewi and his family and the collection of his documents may explain this (F.L.). The selection of the reprinted articles of different authors seems to be justified, as they reflect their specific views.

In comparison, the life and work of Dale are so well documented by Feldberg that no further comments are necessary. The same is true for the documentation of Feldberg.

There is no doubt that many others could have been mentioned: the work of George Koelle, H. Blaschko, Marthe Vogt, J.H. Gaddum, Ulf von Euler, Peter Holtz, D. Nachmannsohn and B. Katz, W.D.M. Paton and Elenor Zaimis, for example. Essential contributions by European, American and Japanese colleagues demonstrate how widely research on neurotransmission has spread. To include all this, there would have been enough information to fill a handbook. In this regard, this book remains incomplete.

Not only scientists, but also substances have their history. Their origin in nature, their isolation and synthesis, and the resulting analogues for therapeutic applications will be brought back to mind.

In modern textbooks of pharmacology, the history of today's drugs is no longer mentioned, due to a lack of space. In the following part of the book the early discoveries of neurotransmitter substances are described. Neurotransmission is the function of a substance released from a nerve terminal to act on one or several different receptors. Almost all neurotransmitters, except peptides, occur also in quite different localizations and serve quite different or even unkown functions. As an example, acetylcholine, when released from nerves, acts on muscarinic or nicotinic receptors, but it is also formed by bacteria and in plants. Cholinergic agonists, like musacrine, nicotine or pilocarpine are formed in plants. Antagonists of acetylcholine, like atropine or d-tubocurarine, are produced in plants where their functions are unkown. Similar examples could be mentioned for adrenaline, histamine or serotonin.

A Substance Isolated from Brain, Synthesized and Called Neurin: Acetylcholine

A historical overview on acetylcholine should not only deal with the substance and its role as a neurotransmitter itself, but should also include all the naturally occurring alkaloids known to mimic or prolong the action of acetylcholine, or to influence its action as antagonists.

The first isolation of acteylcholine from brain was reported in

1864: Oscar Liebreich [1864] prepares a lipid-free extract from brain which he calls 'protagon'.

By alkaline extraction of protagon, 'cleavage products' were found: glycerol phosphoric acid and a basic compound which was named 'neurin'. Adolf von Bayer [1866, 1867] crystallized neurin, identified it as acetylcholine and was able to synthesize it. He was awarded the Nobel Prize for Chemistry for the synthesis of indol, eosin and indigo in 1905. Between the time of its synthesis and the characterization of acetylcholine as the 'Vagusstoff', a number of related findings paved the way [for reviews, see Burgen, 1993, 1995].

A chronological listing of the important events follows, starting already before the year 1864, when 'neurin' was isolated. The first part focuses on the exploration of acetylcholine in the autonomic nervous system; the exploration of acetylcholine at the neuromuscular junction follows in the second part.

1855: Dramatic events are described by Christison [1855] following the self-administration of Calabar beans, the source of physostigmine.

This was followed by an investigation of the actions of Calabar bean extracts by Sir Thomas R. Fraser [1863]. He showed that atropine prevented their lethal effect and recommended its use in ophthalmology.

1864: Jobst and Hesse [1864] isolate the alkaloid physostigmine (also called eserine).

Harnack and Witkowski [1876] were the first to study the pharmacology of physostigmine. Loewi and Mansfeld [1911] described its pharmacological actions. Physostigmine was used therapeutically in the treatment of glaucoma. For therapeutic purposes it was soon replaced by the synthetic analogue neostigmine.

The essential property of physostigmine, inhibition of the acetylcholine-splitting cholinesterase, was discovered by Otto Loewi and Ernst Navratil [1926]. The action of acetylcholine, either injected or released by vagus nerve stimulation, was increased and prolonged under physostigmine. In addition, a physiological reflex mechanism was investigated; stimulation of the retina by light induces a reflex by which the sphincter muscle of the pupillae is constricted. The contraction of the sphincter pupillae, resulting from the neurogenic release of acetylcholine, was increased by physostigmine. In the presence of physostigmine the concentration of acetylcholine in the fluid of the anterior chamber of the eye was increased when pupillary constriction was induced by light [Loewi and Engelhart, 1930; Engelhart, 1931]. It has been known for a long time that atropine prevents the pupillary constriction by light causing pupillary dilatation.

Physostigmine served as a key to investigate the role of acetylcholine in the neurotransmission in ganglia and motor end plates (see below).

The dramatic development of organophosphorus anticholinesterases, used either as insecticides or war gases, was been described by Bo Holmstedt [1963]. We learned from intoxications by these substances that acetylcholine is a highly toxic compound. When its rapid destruction by cholinesterase is completely blocked, death occurs by accumulation of acetylcholine within less than 1 h [for a review, see Koelle, 1963].

1865: Atropine, the alkaloid found in Atropa belladonna *and some other Solanaceae, which has been used in medicine since Babylonian times, is isolated and its chemical structure established by Kraut [1868].*

Rudolf Buchheim [1876] described the pharmacological effects of atropine on the iris and the heart. The mode of action of atropine was elucidated by Loewi and Navratil [1924].

1869: Oswald Schmiedeberg and Richard Koppe [1869] describe the action of muscarine, the toxic alkaloid of the mushroom Amanita muscaria.

The researchers pointed out that muscarine and electrical stimulation of the vagus nerve affected the heart in the same fashion and that both effects were antagonized by atropine.

1878: Pilocarpine is described by John N. Langley [1878].

He observed that pilocarpine was 300 times less potent than acetylcholine in lowering blood pressure and 100 times less potent than acetylcholine in stimulating intestinal muscle. Its action was antagonized by atropine.

1889: John N. Langley and W.L. Dickinson [1889] find that nicotine, when painted on a ganglion or injected intravenously, first facilitates and then blocks synaptic transmission.

Nicotine easily penetrates into the ganglionic synapses. By interrupting their transmission the stimulation of fibres entering the ganglion is prevented. Langley made a very thorough investigation of the distribution of pre- and postganglionic fibres. The term 'autonomic nervous system' was introduced by him, as well as the names 'sympathetic' for the thoracolumbal outflow and 'parasympathetic' for the cranial and sacral outflows [for a review, see Fletcher, 1926].

In 1905, Langley [1905] wrote the following prophetic words:

'I conclude then that in all cells two constituents at least must be distinguished, (1) a substance concerned with carrying out the chief functions of the cells, such as contraction, secretion, the formation of metabolic products, and (2) receptive substances especially liable to change and capable of setting the chief substance in action. Further, that nicotine, curari, atropine, pilocarpine, strychnine, and most other alkaloids, as well as the effective material of internal secretions produce their effects by combining with the receptive substance, and not by an action on axon-endings if these are present, nor by a direct action on the chief substance.'

It is interesting to note that Langley [1903], in his original concept of the autonomic nervous system, assumed that autonomic afferents would be 'those fibres which give rise to reflexes in autonomic tissues, and which are incapable of giving rise to sensation'. The lack of morphological criteria, and the fact that osmium did not stain unmyelinated visceral afferents, led Langley to exclude the existence of autonomic afferent nerves in his further publications and to introduce his 'efferent only version' of the autonomic nervous system. The afferent part of the autonomic nervous system was neglected, although it was known that all morphologically defined autonomic nerves contained a large number of afferent fibres in addition to the cholinergic and adrenergic efferent fibres. In the vagus nerve, 90% of the fibres are afferents, and in the splanchnic nerve afferent fibres make up 50% of the total number of fibres [Goodman and Gilman, 1941]. Morphological markers for unmyelinated afferent fibres were not available until after the discovery of peptidergic afferent neurons from 1965 on. Langley also made a clear distinction between the 'autonomic nervous system' and the 'intrinsic enteric nervous system', the latter being a different, fully autonomic entity, exemplified by the peristaltic reflex [see Holzer and Holzer-Petsche, 1997a, b]. It comprises a nervous system containing as many neurons as the spinal cord. Although autonomic per se, it is modulated by efferent fibres of the autonomic nervous system.

1899: Reid Hunt [1899] notes that adrenal extracts, from which adrenaline has been removed, cause a fall in blood pressure.

Although Hunt identified the presence of choline in the extract, he suggested that some precursor of choline, perhaps an ester, was the vasodepressor agent. Hunt and Taveau [1906] showed that acetylcholine was 100,000 times more active than choline in causing a fall in blood pressure.

1907: Walther Straub [1907] describes the chemical kinetics of the action of muscarine and of the antagonism between muscarine and atropine.

1907: Walter E. Dixon [1907; quoted by Gunn, 1932] concludes from ingeniously designed experiments that an inhibitory substance is released from vagal nerve endings in the heart.

In freshly extirpated frog hearts Dixon stimulated the vagus nerve for half an hour, thereafter placed the heart for 10 s in boiling water (destroying cholinesterase!) and made an alcoholic extract of the heart which was evaporated to dryness. The powder was dissolved in saline solution and showed an inhibition of contractions of other isolated frog hearts. This action, like that of muscarine, was blocked by atropine. Control extracts from hearts whose vagus nerves were unstimulated showed only a much smaller inhibitory effect. Muscarine seemed to be responsible for the action of vagus stimulation on the heart. Dixon wrote: 'I interpret these experiments to mean that some inhibitory substance is stored up in that portion of the heart to which we refer as a "nerve ending", that when the vagus portion is excited this inhibitory substance is set free, and by combining with a body in the cardiac muscle brings about the inhibition. If cardiac inhibition is brought about this way, drugs must act by liberating the inhibitory hormone. Atropine either prevents the liberation of the hormone, or saturates the substance in the end organ upon which it acts.'

1914: Henry H. Dale [1914a, b] examines the pharmacological properties of acetylcholine.

Dale knew that acetylcholine, like muscarine, mimicked the effects of electrical stimulation of parasympathetic nerves. Through experiments on changes of blood pressure in spinal cats, a preparation excluding any central nervous actions, he showed that the fall of blood pressure caused by acetylcholine, indicating vasodilatation, was blocked by atropine. Dale named it the 'muscarinic' action. Following atropine, a much higher dose of acetylcholine induced a rise in blood pressure comparable to the ganglionic stimulation by nicotine. Following desensitization of the ganglia with very high doses of nicotine neither nicotine nor acetylcholine showed an effect on the blood pressure. The rise of blood pressure by acetylcholine could, therefore, be explained as an effect on sympathetic ganglia after paralysis of the post-ganglionic parasympathetic effects by atropine. Dale called this the 'nicotinic' action of acetylcholine. This led to the recognition of two different acetylcholine receptors, the 'muscarinic' type on glands, and smooth and cardiac muscle cells, the 'nicotinic' type on ganglion cells and skeletal muscle [Dale, 1914a, b]. Atropine served as a tool to block the former, nicotine as the tool to first stimulate and then inhibit the latter type of acetylcholine receptors.

1917: Hermann Fühner [1917] finds that the effect of acetylcholine can be enhanced by physostigmine.

Fühner observed that the weak stimulatory effect of acetylcholine on the isolated leech muscle could be enhanced a millionfold by physostigmine at concentrations which were solely response-enhancing but not active on their own, while the effect of choline was not enhanced. He proposed a straightforward model of the mechanism of action: 'This extraordinary potentiation appears to have a simple chemical explanation' provided one considers 'that acetylcholine ... is split into choline and acetic acid at its site of action; the degradation is prevented by physostigmine'. On the isolated frog heart, Fühner [1917] found acetylcholine to be 100,000 times more potent than choline; only much later did he realize that a few years earlier Reid Hunt [1915] had already used the same preparation.

The eserinized leech muscle was the essential key for the following investigation of the 'nicotinic actions' of acetylcholine. Wilhelm Feldberg [1977] wrote: 'It was the use of eserine, its intravenous injection or, in perfusion experiments, its addition to the perfusion fluid in order to inhibit the enzymatic destruction of the released acetylcholine and, to detect it in the venous effluent, blood or perfusate, by means of a simple, specific and sensitive test, the eserinized leech muscle preparation.' When Feldberg arrived in London in 1933, Dale had almost stopped working, but now he was enjoying doing experiments again. Twenty-four papers on ganglionic and neuromuscular transmission by acetylcholine were published by Dale, Feldberg and others in the following three years. Feldberg and Vartiainen [1934] finally were able to estimate the number of molecules of acetylcholine released by a single preganglionic volley per synapse – about 3 million molecules per ganglion!

1920–1932: Alfred J. Clark [1926, 1985] uses acetylcholine as a specific pharmacological tool for his quantitative studies of drug receptor interaction.

Clark wrote that pharmacology may be defined 'as the study of the manner in which the function of living organisms can be modified by chemical substances'. He further stated that the actions of such drugs must be interpreted 'by the known laws of physical chemistry'. He concluded 'that a reversible monomolecular reaction occurs between the drug and some substance either in the cell or on its surface'. He reached this conclusion by experiments on a strip of the ventricle of the frog's heart: when the heart was stimulated electrically at 3-second intervals, the contraction was reduced by acetylcholine. 'Acetylcholine is exceptionally well adapted for the study of laws governing the reactions between drugs and cells for the following reasons:

(1) it produces graded reactions over a wide range of concentrations,

(2) it affects a variety of tissues and produces different types of actions in different tissues,

(3) it acts rapidly and its action is quickly and completely reversible, hence a large number of experiments can be performed on a single preparation' [Clark, 1926].

Clark calculated that the amount of acetylcholine involved in the reaction was too small to cover the whole cell surface in a one-molecule-thick layer. He derived a calculation in agreement with the law of mass action, providing a basis for the quantitative relationship between dose x and effect y:

$$k \cdot x = y/(100 - y)$$

'This relation suggests that a reversible monomolecular reaction occurs between the drug and some substance either in the cell or on its surface.' Using acetylcholine and atropine, he established the quantitative relationships expressed in the equation:

$$K \text{ (conc. ACh)}/\text{(conc. Atr)} = y/(100 - y)$$

He also calculated that the quantity of drug bound by the cell is very small. For a demonstrable action 20,000 molecules per cell are sufficient; these could occupy only a very small fraction of its surface [for references, see Clark, 1985].

1921–1938: Otto Loewi provides the evidence that the action of nerve stimulation is based on the release of acetylcholine from nerve terminals.

In 1921 Otto Loewi [1921] used a pair of isolated frog hearts, the first one with the nerves attached, the second one to be stimulated by the neurotransmitter released into the ventricle of the first heart. Already in 1897 Loewi [1897] had used the isolated frog heart when working on his thesis in Schmiedeberg's laboratory and used it for all his further investigations on neurotransmission. This 'humoral transmission of nervous impulses' was first shown with vagus nerve stimulation and the transmitter was called the Vagusstoff. As some readers may not be familiar with Loewi's most important paper, I would like to mention some details. The paper is four pages long and includes five figures, the results are introduced by the sentence 'Sämtliche Versuche fielen gleichsinnig aus' ('all experiments gave the same results'), thus indicating that biostatistics were not yet available. There is no list of references. Only after a considerable number of further experiments did Loewi [1926] decide that there was sufficient evidence to regard acetylcholine as the Vagusstoff: 'In view of the extremely high activity of the vagal substance it might be thought that it is acetylcholine.' Dale [1914a, b] also considered acetylcholine as a possible neurotransmitter, but hesitated to express this, since acetylcholine had so far not been found in nerves. It was not until 1935 that he wrote [Dale, 1935]: 'Taken together, the evidence is now very strong that the peripheral transmitter involved in responses to parasympathetic nerve stimulation is induced by acetylcholine, and it does not seem to be any longer necessary to speak of an acetylcholine-like substance.'

Loewi's findings of the release of acetylcholine in the frog heart were confirmed in the mammalian heart [Feldberg and Krayer, 1933; for a review, see Whittacker, 1988]; its role in ganglionic transmission was reviewed by Kharkevich [1980], Waser [1986] and Clementi et al. [2000]. Dale [1937] first showed the role of acetylcholine at the neuromuscular junction [for a review, see Zaimis, 1976]. The first nicotinic synapse investigated in human brain was that in the supraoptic nucleus [for references, see Burn, 1948]. The multiple cholinergic pathways in the central nervous system, the distribution of nicotinic and muscarinic receptors in the central nervous system and the connections with other transmitter systems cover chapters in modern textbooks of pharmacology [see e.g. chap. 31 and 33 in Rang et al., 2003].

In recent years, molecular biological methods made possible the isolation and characterization of the structure of muscarinic and nicotinic acetylcholine receptors as documented by the *IUPHAR Compendium of Receptor Characterization and Classification* [2000; muscarinic receptors pp. 54–63, nicotinic receptors pp. 64–77]. Muscarinic receptors exist on the target cells of parasympathetic nerves, nicotinic receptors in autonomic ganglia and on striated muscle; both receptor types occur in the brain and on nerve terminals. Muscarinic receptors convey cellular actions via a G-protein-linked process which lasts seconds, nicotinic receptors are coupled to an ion channel that is activated within milliseconds. The receptor molecules are proteins which are more than a hundred times larger than their agonist. They are synthesized by the cells that express them. We must no longer think of the receptors as fixed elements in cellular control systems that only respond to changes in the concentration of ligands. Receptors themselves are subject to adaptive processes, like upregulation following denervation, alterations due to inflammation, or adaptive responses to the continuous presence of drugs, specifically in the central nervous system.

It might be noted that Loewi and Hellauer [1938a, b] found that the optic nerve and the dorsal roots of the spinal cord, which contain only afferent fibres, are free of acetylcholine. It was not until 1953 that Lembeck found the high amount of substance P in dorsal roots, indicating its role in afferent neurotransmission which was confirmed by Konishi and Otsuka in 1974 [for a review, see Pernow, 1983].

The exploration of the neuromuscular transmission of impulses deserves special attention.

1857: Claude Bernard [1857] shows that curare paralyses the muscle by an effect on the motor end plate.

With ingeniously designed simple experiments on the innervation of the frog gastrocnemius muscle Bernard proved that curare did not abolish nerve conduction or muscle contractility but paralyzed skeletal muscle by acting on the motor end plate. The site of action of curare was, therefore, defined 80 years earlier than its mode of action [Dale et al., 1935; Dale, 1937; for a review of further work, see Zaimis, 1976].

1877: Emil du Bois-Reymond already suggests a chemical transmission from motor nerve endings to the striated muscle [quoted by Dale, 1937].

Thomas R. Elliott is the first to conceive the idea of chemical transmission in the neuromuscular junction.

Elliott tried to extract the active substance from muscle plates of striated muscles, as did Herring from the electric organ of the skate in 1914 but without success. Feldberg [1977] explained this failure: 'They had simply used water or saline extracts. If they only had used hydrochloric acid and thus prevented the spontaneous and enzymic hydrolysis of acetylcholine, they could hardly have missed the high activity in the extracts from the electric organ.'

At the neuromuscular junction, which has been studied more extensively than any other cholinergic synapse, a single nerve impulse releases about 300 synaptic vesicles containing acetylcholine in a concentration of about 100 mmol/l, altogether about 3 million molecules of acetylcholine, within 1 ms from the nerve terminals supplying a single muscle fibre. About 30 million molecules of acetylcholine combine with receptors on each muscle fibre, remain bound to the receptor for about 2 ms, and thereafter are quickly hydrolyzed by the acetylcholinesterase present at the basement membrane between the pre- and postsynaptic membranes. This results in a very rapid and very brief transmission process, being completed within about 20 ms. For an illustration of the speed, it can be compared with that in a cylinder of a combustion engine running at the speed of 3,000 rpm.

Animals with an electric organ use a serial assembly of hundreds of structures comparable to the neuromuscular junction. Simultaneous release of acetylcholine in such an electric 'power station' creates a very high voltage for the animals' defence.

Fessard and Feldberg investigated the role of acetylcholine in the electric organ of *Torpedo marmorata*. In the course of their study, they finally wanted to demonstrate the electrogenic property of acetylcholine injected intra-arterially into the electric organ. So the fish and Feldberg were imprisoned in a small screened cubicle. The sensitive recording instruments, connected by a shielded cable with the electric organ of the fish, were outside the chamber. The injection of acetylcholine by Feldberg resulted in an angry shout from the outside. The electrical response had nearly wrecked the apparatus. What an explosive discovery of the action of acetylcholine! [according to Feldberg, 1977].

The methonium group in choline led Barlow and Ing [1947] to synthetize symmetrical bisquaternary ammonium compounds in which these groups were connected by a polymethylene bridge of varying length. The hexamethylene compound showed ganglionic blocking action, the decamethylene compound, decamethonium, affected the neuromuscular junction. This first 'depolarizing muscle relaxant' was followed by suxamethonium, which has the methonium groups at the same distance as decamethonium. Suxamethonium is still in clinical use for muscle relaxation [Paton and Zaimis, 1852; for a review, see Zaimis, 1976; Clementi et al., 2000].

It is much less known that acetylcholine is found in places not related to neurotransmission. Thus, acetylcholine is present in the nerve-free human placenta [Loewi and Navratil, 1926; Wessler et al., 2001; for a review, see Grando et al., 2003].

The 'burning stings' (emergencies) on the leaves of the nettle *(Urtica dioica)* consist of fine capillary tubes (10 μm diameter), calcified at their lower and silicated at their upper ends and closed at the tip in the form of a little bulb. This bulb breaks off at a predetermined line (like that of ampoules used in medicine) when it comes into contact with the skin, leaving exposed a fine needle-like point formed by the upper tapering part of the hair. The pressure of the contact with the skin makes this fine tube penetrate the skin, and the compression of the bladder-like base injects the fluid into the minute wound. The stings of the nettle contain >1% acetylcholine and >0.1% histamine [Emmelin and Feldberg, 1947] in addition to serotonin (4 ng/sting) [Collier and Chesher, 1956]. The stings, therefore, contain a 'pandemonium' of compounds stimulating afferent pain fibres [Lembeck and Juan, 1974] – which is a pharmacological explanation for the well-known urticaria and pain following the contact with leaves of *U. dioica*!

Dale [1914a] found acetylcholine in a few, but not all, ergot alkaloid-containing extracts of *Claviceps purpurea*. However, this was caused by bacterial contamination.

Acetylcholine is produced by a choline acetyltransferase in cultures of *Lactobacillus arabinosus*, like that of sauerkraut [Lembeck and Schraven, 1964]. This shows that acetylcholine existed already very early in evolution.

Finally, it has to be mentioned that quite a number of substances found elsewhere in nature or synthesized compounds are known to act on acetylcholine receptors or by inhibition of acetylcholinesterase. On muscarinic receptors, agonistic action is evoked by muscarine, the long-acting pilocarpine and bethanechol (not split by cholinesterase); antagonistic actions are seen with hyoscine, atropine and scopolamine and their synthetic analogues. On nicotinic receptors agonistic actions are known not only of nicotine and the much weaker-acting lobeline, but also of the recently discovered epibatidine from the skin of an Ecuadorian frog with a 100-fold higher potency than nicotine; antagonistic actions are observed after high doses of nicotine, as well as after hexamethonium and trimethaphan.

On the neuromuscular junction, it has to be differentiated between muscles which respond to depolarization by long-lasting contractions, like the bird muscles and the extra-ocular muscles in mammals, and those which do not, like the diaphragm after stimulation of the phrenic nerve [Bülbring, 1946]. In the first ones, epibatidine is a 100 times more potent that suxamethonium or nicotine, and a 1,000 times more potent than acetylcholine [Lembeck, 1999]. Epibatidine, once used as an arrow poison, is therefore the most potent substance acting on neuromuscular nicotinic acetylcholine receptors. An antagonistic effect on the neuromuscular junction is known to exist with *d*-tubocurarine, pancuronium and similar synthetic compounds.

Botulinum toxin is a bacterial exotoxin produced by *Clostridium botulinum* under anaerobic conditions in preserved food. In the mouse it has a minimal lethal dose of 10^{-12} g, i.e. a few million molecules. Botulinum toxin consists of peptidases which cleave specific proteins involved in exocytosis, thereby producing a long-lasting block predominantly in parasympathetic and motor functions. Botulinum toxin, injected locally into muscles, is used to treat a form of eye lid spasm, as well as other types of spasticity [Rang et al., 2003].

Physostigmine (eserine) and neostigmine, the analogues most frequently used in therapy, block acetylcholinesterase in a reversible manner, whereas organo-phosphorus compounds are irreversible inhibitors of the enzyme.

Myasthenia gravis is a disease based on an auto-immune response which destroys the function of the nicotinic acetylcholine receptors at the neuromuscular junction. It results in the inability of muscles to produce sustained contraction, a characteristic sign of which are drooping eyelids. The first attempt of a treatment with physostigmine was made by Walker [1934], long before the cause of the disease was known.

Recent investigations centred around the involvement of the cholinergic system of the brain in Alzheimer's disease and in the related dementia of Down's syndrome. A pathological cascade in the disease-related memory loss could be triggered by alterations in acetylcholine-mediated neuronal functions [Isacson et al., 2002]. A therapeutic trial with anticholinesterases is therefore tempting. Combinatorial chemistry might offer a new approach towards developing specific compounds. Certain inhibitors of acetylcholinesterase dock on two different sites of the large cholinesterase molecule; once bound, the two compounds react with each other to form a highly effective anticholinesterase [Lewis et al., 2002].

Through the experimental use of acetylcholine an important discovery was made in recent years. Robert Furchgott and Zawidzki [1958] investigated the relaxation of vascular smooth muscles using rings of rabbit aorta precontracted with noradrenaline. Under these conditions the rings relaxed following the addition of acetylcholine provided that the handling of the ring preparation had been done very carefully. If the vessels had been rubbed between the fingers, the rings failed to relax in response to acetylcholine. Such a handling removed the thin sheet of endothelium and Furchgott concluded that acetylcholine released an 'endothelium-derived relaxing factor' (EDRF). Subsequent work by Salvador Moncada, Louis Ignarro and many others revealed that EDRF acts only for seconds. The structure of this compound was a great surprise: it is nitric oxide, NO (see pp. 201/202). The rapidly increasing work on this topic showed the manyfold significance of NO and led to the 1999 Nobel Prize being awarded to Robert Furchgott, Ferid Murad and Louis Ignarro.

The endothelial *L*-arginine/NO pathway is tonically active in resistance vessels, providing a physiological vasodilator mechanism. The 'muscarinic' vasodilator action of acetylcholine, measured by Dale as early as 1914 as an immediate fall in blood pressure, was interpreted by Furchgott in his experiments in 1958 with NO as mediator.

In conclusion, it is perhaps worth mentioning that amongst all the many highly potent endogenous compounds only acetylcholine and NO are known by their chemical names, and do not have 'catchy' names such as serotonin, bradykinin and other terms.

References

Barlow R, Ing HR: Polymethylene bis-quaternary ammonium salts. Br J Pharmacol 1947;3:294–304.

Bayer A: Synthese des Neurins. Ann Chem Pharm 1866;140: 306–313.

Bayer A: Über das Neurin. Ann Chem Pharm 1867;58: 322–326.

Bernard C: Vingt-deuxième leçon: action du curare. Leçons sur les effets des substances toxiques et médicamenteuses. Paris, Baillières, 1857, pp 326–327.

Buchheim R: Über die pharmakologische Gruppe des Atropins. Arch Exp Pathol Pharmakol 1876;5:463–472.

Bülbring E: Observations on the isolated phrenic nerve diaphragm preparation of the rat. Br J Pharmacol 1946;1: 38–61.

Burgen AS: Evolution and acetylcholine receptors. Prog Brain Res 1993;98:129–131.

Burgen AS: The background of the muscarinic system. Life Sci 1995;56:801–806.

Burn JH: The hypothalamus as reflex center; in The Background of Therapeutics. London, Oxford University Press, 1948, pp 14–35.

Christison R: On the properties of the ordeal bean of Old Calabar, Western Africa. Pharm J 1855;14:470.

Clark AJ: The antagonism of acetylcholine by atropine. J Physiol 1926;64:547.

Clark DH: Alfred Joseph Clark 1885–1941; a Memoir. C & J Clark Ltd. Archives, 1985.

Clementi F, Fornasari D, Gotti C: Neuronal nicotinic receptors. Handb Exp Pharmacol. Berlin, Springer, 2000, vol 144.

Collier HOJ, Chesher GB: Identification of 5-hydroxytryptamine in the sting of the nettle *(Urtica dioica)*. Br J Pharmacol 1956;11:186–189.

Dale HH: The action of certain esters and ethers of choline, and their relation to muscarine. J Pharmacol 1914a;6: 147–190.

Dale HH: The occurrence in ergot and action of acetylcholine. J Physiol 1914b;48:iii.

Dale HH: Walter Ernest Dixon Memorial Lecture. Proc R Soc Med Ther Sect 1935;28:319.

Dale HH: Du Bois-Reymond and the chemical transmission. J Physiol 1937;91:4P.

Dale HH, Feldberg W, Vogt M: Release of acetylcholine at voluntary motor nerve endings. J Physiol 1935;86: 353–380.

Dixon WE: On the mode of action of drugs. Med Magazin (Lond) 1907;16:454–457.

Emmelin N, Feldberg W: The mechanism of the sting of the common nettle *(Urtica dioica)*. J Physiol 1947;106: 440.

Engelhart E: Der humorale Wirkungsmechanismus der Oculomotoriusreizung. Pflügers Arch Physiol 1931;227: 220–234.

Feldberg W: The Early History of Synaptic and Neuromuscular Transmission by Acetylcholine: Reminiscence of an Eye Witness; in The Pursuit of Nature. London, Cambridge University Press, 1977, pp 65–83.

Feldberg W, Krayer O: Das Auftreten eines azetylcholinartigen Stoffes im Herzvenenblut vom Warmblütern bei Reizung der Nervi vagi. Arch Exp Pathol Pharmakol 1933;172:170–193.

Feldberg W, Vartiainen A: Further observations on the physiology and pharmacology of a sympathetic ganglion. J Physiol 1934;83:103–128.

Fletcher WM: John Newport Langley. In memoriam. J Physiol 1926;61:1–27.

Fraser TR: On the characters, actions and therapeutical uses of the ordeal bean of Calabar. Edinburgh Med J 1863;9: 36–56, 123–132, 235–248.

Führer H: Ein Vorlesungsversuch zur Demonstration der erregbarkeits-steigernden Wirkung des Physostigmins. Arch Exp Pathol Pharmakol 1917;42:81–85.

Furchgott RF, Zawidzki JV: The obligatory role of endothelial cells in the relaxation of arterial smooth muscle by acetylcholine. Nature 1958;288:373–376.

Goodman L, Gilman A: The Pharmacological Basis of Therapeutics, ed 1. New York, Macmillan, 1941, p 318.

Grando SA, Kawashima K, Wessler I: Non-neural acetylcholine. Signalling pathways and major biological effects in humans. Life Sci 2003;72:2009–2181.

Gunn JA: Walter Ernest Dixon 1871–1931. J Pharmacol 1932; 44:1–21.

Harnack E, Witkowski L: Pharmakologische Untersuchungen über das Physostigmin und Calabarin. Arch Exp Pathol Pharmakol 1876;5:401–454.

Holmstedt B: Structure-activity relationships of the organophosphorus anticholinesterase agents. Handb Exp Pharmacol. Berlin, Springer, 1963, vol 15, pp 428–485.

Holzer P, Holzer-Petsche U: Tachykinins in the gut. I. Expression, release and motor functions. Pharmacol Ther 1997a;73:173–217.

Holzer P, Holzer-Petsche U: Tachykinins in the gut. II. Roles in neural excitation, secretion and inflammation. Pharmacol Ther 1997b;73:219–263.

Hunt R: Note on a blood-pressure lowering body in the suprarenal gland. Am J Physiol 1899;3:XVIII–XIX.

Hunt R: A physiological test for choline and some of its applications. J Pharmacol Ther 1915;7:301–337.

Hunt R, Taveau M: On the physiological action of certain choline derivatives and a new method for detecting choline. Br Med J 1906;II:1788–1789.

Isacson O, Seo H, Lin L, Albeck D, Granholm AC: Alzheimer's disease and Down's syndome: roles of APP, trophic factors and ACh. Trends Neurosci 2002;25:79–84.

IUPHAR Compendium of Receptor Characterization and Classification. London, IUPHAR Media, 2000.

Jobst J, Hesse O: Physostigmin. Am J Pharmacol 1864, pp 334, 365.

Kharkevich A: Pharmacology of ganglionic transmission. Handb Exp Pharmacol. Berlin, Springer, 1980, vol 54.

Koelle GB: Cholinesterases and Anticholinesterases. Handb Exp Pharmacol. Berlin, Springer, 1963, vol 15.

Kraut K: Die chemische Struktur von Atropin. Liebigs Ann 1868;148:236.

Langley JN: The action of pilocarpin on the submaxillary gland of the dog. J Anat Physiol 1876;11:173–180.

Langley JN: The autonomic nervous system. Brain 1903;26: 1–26.

Langley JN: On the reaction of cells and of nerve endings to certain poisons, chiefly as regards the reaction of striated muscle to nicotine and to curare. J Physiol 1905; 33:374–413.

Langley JN, Dickinson WL: On the local paralysis of peripheral ganglia, and on the connection of different classes of nerve fibres with them. Proc R Soc 1889;46:423–431.

Lembeck F: Epibatidine: high potency and broad spectrum activity on neuronal and neuromuscular nicotinic acetylcholine receptors. Naunyn Schmiedebergs Arch Pharmacol 1999;339:378–385.

Lembeck F, Juan H: Interaction of prostaglandins and indomethacin with algesic substances. Naunyn Schmiedebergs Arch Pharmacol 1974;285:301–313.

Lembeck F, Schraven: Der Einfluss von Hemicholinium Nr. 3 auf die Acetylcholin-Bildung in *Lactobacillus arabinosus*. Naunyn Schmiedebergs Arch Exp Pathol Pharmacol 1964;247:93–99.

Lewis WG, Green LG, Grynspan F, Radic Z, Carlier PR, Taylor P, Finn MG, Sharpless KB: Click chemistry in situ: acetyllcholinesterase as a reaction vessel for the selective assembly of a fentomolar inhibitor from an array of building blocks. Angew Chemie 2002;114:1095–1099.

Liebreich O: Über die chemische Beschaffenheit der Gehirnsubstanz. Ann Chem Pharm 1864;54:29–44

Loewi O: Zur quantitativen Wirkung von Blausäure, Arsen und Phosphor auf das isolierte Froschherz. Arch Exp Pathol Pharmacol 1897;38:127–138.

Loewi O: Über die humorale Übertragbarkeit der Herznervenwirkung. Pflügers Arch Ges Physiol 1921;189: 239–242.

Loewi O: Über die Wirkung von Muskarin auf das nicht oder unzureichend gespeiste Froschherz und die Gegenwirkung von Kaliumsalzen. Zentralbl Physiol 1926; 19:593–595.

Loewi O, Engelhart E: Fermentative Azetylcholinspaltung im Blut und ihre Hemmung durch Physostigmin. Arch Exp Pathol Pharmakol 1930;150:1–13.

Loewi O, Hellauer H: The acetylcholine content of the nerves in warm-blooded animals. J Physiol 1938a;93:34P.

Loewi O, Hellauer H: Über das Acetylcholin im peripheren Nerven. Pflügers Arch Ges Physiol 1938b;240:769–775.

Loewi O, Mansfeld G: Über den Wirkungsmodus des Physostigmins. Arch Exp Pathol Pharmakol 1911;62:180–185.

Loewi O, Navratil E: Über humorale Übertragbarkeit der Herznervenwirkung: der Angriffspunkt des Atropins. Pflügers Arch Ges Physiol 1924;206:123–134.
Loewi O, Navratil E: Über den Mechanismus der Vaguswirkung von Physostigmin und Ergotamin. Pflügers Arch Ges Physiol 1926;214:688–696.
Paton WDM, Zaimis E: The methonium compounds. Pharmacol Rev 1952;4:219–253.
Pernow B: Substance P. Pharmacol Rev 1983;35:85–141.
Rang HP, Dale MM, Ritter JM, Moore PK: Pharmacology, ed 5. Edinburgh, Churchill Livingstone, 2003.
Schmiedeberg O, Koppe R: Das Muscarin. Das giftige Alkaloid des Fliegenpilzes. Leipzig, Vogel, 1869.
Straub W: Zur chemischen Kinetik der Muskarinwirkung und des Antagonismus Muskarin-Atropin. Pflügers Arch Ges Physiol 1907;119:127–151.
Walker MB: Treatment of myastenia gravis with physostigmine. Lancet 1934;226:1200–1201.
Waser PG: The cholinergic receptor; in Parnham MJ, Bruinvels J (eds): Discoveries in Pharmacology. Amsterdam, Elsevier Science, 1986, vol 3.
Wessler I, Roth E, Schwarze S, Weikel W, Bittinger F, Kirkpatrick CJ, Kilbinger H: Release of non-neuronal acetylcholine from the human placenta: difference to neuronal acetylcholine. Naunyn Schmiedebergs Arch Pharmacol 2001;364:205–212.
Whittacker VP: Cholinergic synapses. Handb Exp Pharmacol. Berlin, Springer, 1988, vol 86.
Zaimis E: Neuromuscular junction. Handb Exp Pharmacol. Berlin, Springer, 1976, vol 42.

Adrenaline, Noradrenaline and Dopamine: The Catecholamines

From the 'Acceleransstoff' to Adrenaline, Noradrenaline and Dopamine

An exciting event was the overture in the discovery of adrenaline, described by Dale [1949]:

'Dr. George Oliver, a physician of Harrogate, employed his winter leisure in experiments on his family, using apparatus of his own devising for clinical measurements. In one such experiment he was applying an instrument for measuring the thickness of the radial artery; and having given his young son, who deserves a special memorial, an injection of an extract of the suprarenal gland, prepared from material supplied by the local butcher. Oliver thought that he detected a contraction or, according to some who have transmitted the story, an expansion of the radial artery. Whichever it was, he went up to London to tell Professor Schäfer what he thought he had observed, and found him engaged in an experiment in which the blood pressure of a dog was being recorded; found him, not unnaturally, incredulous about Oliver's story and very impatient at the interruption. But Oliver was in no hurry, and urged only that a dose of his suprarenal extract, which he produced from his pocket, should be injected into a vein when Schäfer's own experiment was finished. And so, just to convince Oliver that it was all nonsense, Schäfer gave the injection, and then stood amazed to see the mercury mounting in the arterial manometer till the recording float was lifted almost out of the distal limb.

Thus the extremely active substance formed in one part of the suprarenal gland, and known as adrenaline, was discovered. And in due course there came to light the curious correspondence between the effects produced by this potent substance and those produced by nerves of the so-called sympathetic system; and Professor T.R. Elliott, then a postgraduate research student in Cambridge, was led to make the brilliant suggestion that these sympathetic nerves produce their effects by liberating small quantities of adrenaline at the points where they end in contact with muscle fibres and gland cells.'

Following their finding with adrenal gland extracts, Oliver and Schäfer [1895a, b] also found a rise of blood pressure caused by extracts of the pituitary; compared with the extract from the adrenals, its action was slower in onset, maintained for a longer time and there was no accelerating action on the heart.

The structure of adrenaline, its synthesis and the separation of both enantiomers were established within a few years. Barger and Dale

[1910] studied the pharmacology of a large number of synthetic amines related to adrenaline and they termed their action 'sympathomimetic'. Among the substances tested was also noradrenaline, which was already obtainable from Hoechst under the name 'arterenol'. In his comment on the paper, written in 1952, Dale wrote: 'I failed to jump to the truth' and regretted that his friend Walter B. Cannon spent so much time on his hypothesis of two complex 'sympathins', one E (excitatory) and the other one I (inhibitory), to explain the differences between the effects of adrenaline and those caused by sympathetic nerve stimulation. Baqu [1934] suggested that 'sympathin E' might be noradrenaline. Adrenaline is also called epinephrine, and occasionally this has caused some confusion, as described by Aronson [2000].

Otto Loewi [1921, 1924] was very cautious in the interpretation of the nature of substances in the humoral transmission of cardiac innervation. He named the transmitters the 'Vagusstoff' and the 'Acceleransstoff'. Five years later he had sufficient evidence that the former is acetylcholine, and in 1936, that is 15 years after the first findings, by using the chemical method of Gaddum and Schild [1934], he knew that the latter is noradrenaline [Loewi, 1936].

Ulf von Euler [1946] and Peter Holtz et al. [1944/1947] deserve the merit for the conclusive evidence that noradrenaline is the predominant transmitter of sympathetic nerve impulses. The paper of Holtz et al. was 'in print' already in 1943, when it went up in flames during an air raid on Leipzig. This discovery was soon followed by complementary results, such as those of Holtz and Schümann [1948] on the circulation, by the detection of noradrenaline in human urine [Holtz et al., 1944/1947], by observations on the innervated versus denervated nictitating membrane of the cat's eye [Bülbring and Burn, 1949], and by multiple screening experiments [Gaddum et al., 1949] to distinguish between adrenaline and noradrenaline.

It was the time of classical bioassays, which were rather complicated in terms of modern physical or biochemical methods. I would like to mention one example from my personal experience: the isolated rat uterus and colon respond to carbachol (more stable in solution than acetylcholine) with regular reproducible contractions. These contractions can be inhibited by adrenaline and noradrenaline; the sensitivity of the colon is of the same magnitude for both catecholamines, but the uterus is 100 times more sensitive to adrenaline than to noradrenaline. The sensitivity was in the range of nanograms and, therefore, suitable, for instance, to measure both of them in the venous blood from the cat's adrenals. By comparing the effects of a sample of venous blood with that of adrenaline and noradrenaline both on the uterus and the colon, the amount of adrenaline and noradrenaline could be calculated. The additions of carbachol in 2-min intervals and the following washout made great demands on the patience of the investigator. Mr. Condon, the Department's technician, solved the problem; he brought mechanical telephone relays from an army surplus store and constructed an automatically operating device for the addition and washout of carbachol [Gaddum and Lembeck, 1949].

A third endogenous catecholamine turned up: dopamine also had transmitter functions, and was not only a precursor in the synthesis of noradrenaline and adrenaline. In 1954, Marthe Vogt [1954] described what she called 'brain sympathin': it was identified as noradrenaline, present chiefly in hypothalamus and brain stem. A few years later, dopamine was found in the brain [Montague, 1957; Weil-Malherbe and Bone, 1957].

Carlsson [1959] and Carlsson et al. [1962] discovered that dopamine is present in the brains of rabbits in an amount of 0.4 μg/g, a quantity much higher than what would be expected if dopamine were merely a precursor of noradrenaline. Injection of reserpine made dopamine almost completely disappear from the brain, while intravenous injection of the precursor *L*-DOPA caused more than a 4-fold increase of the dopamine present in the brain. Bertler and Rosengren [1959] found in the dog that the caudate nucleus contained about 80% of the total amount of brain dopamine, a finding soon extended to the human brain. The sites with the highest dopamine content contained the smallest amount of noradrenaline. Dopaminergic pathways in the central nervous system were shown to differ from those of noradrenaline, adrenaline, serotonin and acetylcholine [see Rang et al., 2003].

The Impact of New Chemical, Histochemical and Biochemical Methods

Chemical methods made the separation and estimation of all three catecholamines much easier. By paper-chromatographic separation, Goldenberg et al. [1949] surprisingly showed that the 'Epinephrine Standard' of the US Pharmacopoeia, prepared from beef adrenals, was 'contaminated' by up to 36% with noradrenaline. The spectrophotofluorimetry, introduced by Udenfriend [for a review, see 1962], led to a new dimension of sensitivity and specificity of the assays for catecholamines. The next step was the fast separation of catecholamines by HPLC, in combination with electrochemical detection, now used in routine assays in clinical laboratories.

Before 1960, it was known that adrenaline, noradrenaline, dopamine and serotonin were present in peripheral tissues and the brain, but there was only speculation on what kind of cells stored these compounds. Hillarp [1959] succeeded in making visible in the fluorescence microscope the location of catecholamines and serotonin in neurons and endocrine cells in the periphery and in neuronal cell bodies, fibres and nerve terminals in the brain. The use of formaldehyde to induce fluorescence resulted from several methodological components. Upon treatment with formalin vapour, generated from a 35% solution of formaldehyde, fluorescence occurred if the amines were present in a dried protein layer. Proteins catalyzed the reaction of amines to tetrahydro-isoquinolines, and increased temperature enhanced the fluorescence. When cryostate sections were dried in vacuo and exposed to formaldehyde at 50–75 °C, a strong fluorescence developed within 30 s. Noradrenaline was converted to a strongly fluorescent green product, and serotonin to a yellow fluorescence product [Carlsson, 1959; Falck et al., 1962]. Details of this development which had an enormous impact in neuroscience were described by Dahlström and Fuxe [1999] and Roe [1999]. For the first time, the dense network of tiny adrenergic nerve fibres in the iris was made visible by a specific fluorescent colour. What a discovery!

The mapping of serotonin-containing pathways was achieved after the introduction of the immunofluorescence method, using specific antiserotonin antibodies.

Biochemical studies were essential for the study of the formation of catecholamines from tyrosine and its degradation to compounds excreted via the urine. Other than peptides, whose synthesis is stored in the genetic code, only the enzymes, needed for the synthesis and cleavage of catecholamines, are stored in the genetic code.

Three enzymes (tyrosine hydroxylase, DOPA decarboxylase and dopamine β-hydroxylase) are needed for the production of noradrenaline. Only the adrenals and the central nervous system further contain phenylethanolamine N-methyl transferase (PNMT) to form adrenaline [Axelrod, 1962].

The degradation of catecholamines by monoamine oxidase was studied by Blaschko [1952] and Kapeller-Adler [1970]. Axelrod [1957] discovered the enzyme catechol-O-methyl transferase (COMT) which adds an O-methyl group to the catecholamines.

Regulatory Requirements in the Adrenergic System

The function of the adrenergic system depends on three completely different mechanisms: (1) the regulation of the outgrow of adrenergic fibres by nerve growth factor (NGF), (2) the adrenaline synthesis in the adrenals needs, in addition to the three enzymes mentioned before, PNMT and (3) the availability of the precursor, as shown below by the role of levodopa for the formation of dopamine.

(1) Rita Levi-Montalcini and Hamburger [1951] found in 1951 that transplantation of a mouse sarcoma to a chick embryo had an enormous growth-promoting effect on sympathetic and spinal ganglia. The definition of the nature of the factor, of its action and finally of the gene responsible for its production have occupied Levi-Montalcini and many other researchers for almost three decades.

The crucial experiment of Levi-Montalcini was the following: chick embryos were wrapped in the extra-embryonic corio-allantoic membrane which was crossed by a thick network of blood vessels belonging to the embryo itself. The membrane protected the embryo, while permitting gaseous exchange and enabled it to receive nutrients from the egg yolk and to get rid of waste products. She grafted small fragments of the mouse sarcoma close to the chorio-allantoic membrane. After 10 days of incubation she removed the embryo and described the effects which had taken place: sensory ganglia and, even more so, sympathetic ganglia, were found to be much larger in volume. Sympathetic fibres had invaded the organs, some even penetrating into the veins of the host embryo. These results demonstrated irrefutably that an humoral agent of the tumour had stimulated the development of receptive nerve cells.

She learned the technique of in vitro culture of nerve cell lines from Levi. This method had in the meantime been established at the Institute of Biophysics in Rio de Janeiro. A grant from the Rockefeller Foundation enabled her to go to this place. In 1952 on a flight to Brazil, she transported a couple of young mice with the sarcoma in her handbag!

As soon as she had returned to St. Louis, she set up her own in vitro culture of chicken embryos, now jointly with Stanley Cohen. The tumour factor was identified as a nucleoprotein, a macromolecule formed by complexes of nucleic acids and proteins. As it was likely that the nucleic acids could be a contaminant, the fraction was treated with a snake venom, containing the enzyme phosphodiesterase which breaks down nucleic acids. Rita assayed all the fractions including one which contained the snake venom only for control. In one of the fractions, an outstandingly high effect was seen – it was the sample which contained the snake venom only! Cohen purified the fraction of snake venom that was, in terms of relative weight, 3,000 times richer in NGF than the tumour extracts. The accidental discovery that malignant mouse tumours and the venom of snakes share the unique property of promoting the growth of nerve fibres – what an unexpected finding!

The following explorations showed that NGF is synthesized in the submandibular salivary glands of male mice [for more details, see Levi-Montalcini, 1988; Levi-Montalcini and Angeletti, 1968]. Otten et al. [1980] showed that NGF is a trophic factor of unmyelinated afferent neurons. In addition, NGF is a mediator for inflammation [Levi-Montalcini and Skaper, 1996].

(2) The enzyme PNMT, which transforms noradrenaline to adrenaline in the adrenals, is located close to the adrenal cortex. This explains why corticosteroids are able to stimulate the formation of adrenaline to be released from the adrenals. This makes it possible to connect Cannon's [1928] 'emergency reaction' to Selye's 'stress reaction', the former being essentially the adrenergic 'fight and flight reaction', the latter being based on many additional actions of the increased release of glucocorticoids.

PNMT is also found in certain parts of the brain, where there is some evidence that adrenaline may function as a transmitter. In these sites, PNMT is also sensitive to steroid hormones, providing a possible mechanism whereby these hormones can affect brain function.

(3) The supplementation of deficient dopamine in basal ganglia, the problem in Parkinson's patients, by oral application of levodopa is again another story. Birkmayer, a neurologist in Vienna, had the idea that Parkinson's disease might result from an imbalance of neurotransmitters in substantia nigra and striatum. He met the right person at the right time, Oleh Hornykiewicz [1994] who had just made himself familiar with the newest methods for the study of the metabolism of dopamine during 2 years in Blaschko's laboratory in Oxford. Let him tell us the story:

In January 1961, I approached the neurologist Walther Birkmayer with the request to try out a novel pharmacotherapy of Parkinson's disease. I asked Birkmayer to give dopamine's precursor substance levodopa to patients afflicted with this neurodegenerative disorder. I based this dopamine substitution concept on the results of a study in the human postmortem brain that Ehringer and Hornykiewicz [1960] had just before carried out, demonstrating a profound lack of dopamine in the caudate nucleus and putamen of patients with Parkinson's disease. In the Vienna of the early 1960s as in Montreal [for a review, see Sourkes and Gauthier, 1983] the use of levodopa thus had a rational basis right from the beginning. The enormous significance, that this dopamine substitution concept had, was shown 6 years later through the clinical work of Cotzias et al. [1967] in New York. These experiments demonstrated for the first time – and in a very striking manner – the immediate relevance of basic brain research for the diseased patient and marked the beginning of what is now known as modern neuroscience.

Under the infusion of levodopa, the precursor of DOPA, the patient himself could observe how his disturbed handwriting dramatically improved! More important was the fact that levodopa was shown to be orally active. It is the first successful replacement therapy for a disturbed brain function. Now levodopa is administered in combination with a peripherally active inhibitor of aromatic *l*-amino acid decarboxylase, such as carbidopa or benserazide, thus increasing the fraction of levodopa entering the brain and reducing the therapeutic dose to a tenth [for a review, see Hornykiewicz, 1992].

α- and β-Receptors: Agonists and Antagonists

Catecholamines bind to specific receptors where they induce their physiological actions. These adrenoreceptors are characterized and classified as α-, β- and dopamine receptors (including several subtypes of each receptor). Adrenaline acts on α- and β-receptors, noradrenaline more on α- than on β-receptors, and isopropyladrenaline, which does not exist in nature, is the prototype for action on β-receptors.

The localization of the receptors is either on postsynaptic membranes or presynaptically on the terminal which releases the catecholamine. Postsynaptic receptors activate adrenergic responses, and presynaptic α_2-receptors regulate the release of nordrenaline by a feedback mechanism.

Receptors are proteins of several 100-fold the size of the agonist acting on them. They are synthesized by the cells that express them. They cannot be regarded as fixed elements in cellular control systems; they are themselves the subject of regulation. Examples are the proliferation of various postsynaptic receptors after denervation, the upregulation of cytokine receptors in inflammation or the induction of growth receptors.

Dale [1913] provided the first indication of two different adrenergic receptors. Following the injection of the α-receptor blocker ergotoxine (not defined as such at the time), the rise in blood pressure by adrenaline was converted to a decrease in blood pressure, which means, in modern terms, that the β-receptor-mediated action remained. Alquist [1954] clearly differentiated between α- and β-receptors; although he presented solid evidence his paper was rejected by the *Journal of Pharmacological and Experimental Therapy*; it was, however, accepted by the *American Journal of Physiology*. On both receptors, agonistically and antagonistically acting compounds are known.

The prototype for α-adrenergic action is noradrenaline, whereas adrenaline acts on both types of receptors. In terms of physiology, noradrenaline is the neurotransmitter which regulates the distribution of blood at different positions of the body by local vasoconstriction. Adrenaline, on the other hand, released from the adrenals into the vena cava, is distributed by the bloodstream like a hormone; it acts not only on vessels, secretory glands and dilates bronchi, but also influences the carbohydrate and lipid metabolism. In contrast to short-acting vascular effects of intravenously injected adrenaline, the metabolic effects, indicated by the rise of glucose and fatty acids in blood, as well as an increased oxygen consumption, continue for up to 1 h. The reason for the long duration of these effects is that they are mediated by an 'enzymatic clockwork'. This is comparable to a mechanical clockwork where the fast motions of the balance wheel are transformed by a succession of cogwheels in order for the big hand to move just once in an hour.

The short contacts of adrenaline with the receptor activates, via G proteins and cAMP, the induction of increased lipolysis, reduced glycogen synthesis and increased glycogen breakdown to glucose-1-phosphate [Sutherland and Cori, 1951; Sutherland and Rall, 1969].

Companions and Opponents of Catecholamines

Only a few direct therapeutic applications of catecholamines exist: adrenaline in acute anaphylactic reaction with extensive urticaria (type I hypersensitivity), and added to local anaesthetics to prolong their action by vasoconstriction. Several orally active α_1-receptor agonists were temporarily used in orthostatic hypotension. Once, an orally active specific α_1-receptor agonist was found, but its application in hypotension was accompanied by too many unpleasant side effects. Dopamine is used in cardiogenic shock and to increase renal perfusion. In contrast to these few examples of application, a large number of naturally occurring or synthesized compounds of selective actions exists, which are of greater significance.

Ergotamine has an α_1-receptor agonistic action, and is used in the treatment of acute attacks of migraine. Figures in textbooks showed that the large pulse wave in the temporal artery, observed in attacks of migraine, is reduced under ergotamine with simultaneous mitigation of the headache. Under experimental conditions, the α_1-agonistic action of ergotamine is followed by a long-lasting α_1-antagonistic effect.

Loewi's 'Acceleransstoff' [1924] was shown to be inhibited by ergotamine [Loewi and Navratil, 1927]. It is much less known that Dale [1906, 1909] discovered oxytocin in the course of his experiments, which showed the reversal of the rise in blood pressure by adrenaline into a decrease of the blood pressure after ergotoxine. He wanted to define the specificity of this effect by the use of another pressor principle and used a pituitary extract. In contrast to adrenaline, its pressor effect was not inhibited by ergot. He wrote in his comment to the publication: 'I happened in this experiment, to be recording also the tone and rhythm of the cat's uterus, an early pregnant one, and thus by chance, I observed the potent stimulating action on the unstriated muscle of the uterus'; this was the discovery of oxytocin!

Gaddum [1926] devised a bioassay of ergotamine on the uterus of the rabbit stimulated by adrenaline and discovered under ergotamine a parallel shift of the response curve to the right when the concentrations of adrenaline were plotted on a logarithmic scale, which was the first experimental evidence of a competitive inhibition.

Isoprenaline, known today as the prototype of a pure β-agonist, has an interesting background. It was a joint venture of the pharmaceutical company Boehringer Ingelheim and the Department of Pharmacology in Vienna.

Konzett and Rössler [1940] invented a simple apparatus, consisting of a Starling respiration pump, a mercury valve and a piston recorder, to measure bronchoconstriction and bronchodilatation. Bronchoconstriction was induced by vagus nerve stimulation for short, and pilocarpine for long-lasting bronchoconstriction. Konzett [1940a, b] measured the inhibition of induced bronchoconstriction, and found, in comparison with adrenaline, a 10-fold higher potency with isopropyladrenaline (isoprenaline). The compound underwent a careful pharmacological screening, and first studies on humans were conducted on members of the department in Vienna. Boehringer Ingelheim developed an aerosol which has remained the main type of application even today [for details of the discovery, see Konzett, 1981]. Isoprenaline was the first efficient treatment of asthma; the following development has led to therapeutically even more suitable drugs used in asthma therapy today.

Important for all the further experimental work was the finding that isoprenaline was and still is the prototype for a β-agonist. β_2-Selective agonists include all commonly used bronchodilators for asthma. β_1-Selective agonists have only certain indications in cardiology.

Catecholamines are not only synthesized but also stored, along with ATP, in vesicles of the nerve terminals. When a nerve impulse arrives at the terminal, calcium channels are opened, the resulting calcium inflow promotes the fusion and discharge of synaptic vesicles resulting in the exocytotic release of catecholamines, and also ATP, which reach their postsynaptic receptors.

The action of noradrenaline is regulated by a re-uptake mechanism [Iversen, 1971; for a review, see Starke et al., 1989]. The mechanism has been discovered in Axelrod's laboratory in Bethesda. Confusing prior results caused Axelrod to consider the possibility of a catecholamine re-uptake in the nerve terminals, a mechanism unknown elsewhere in the nervous system. How to prove or disprove this re-uptake phenomenon? He and his colleagues exchanged many ideas. Then, one day, he and his visiting scientist Georg Hertting – 'a tiny, little, self-effacing man from Vienna', someone once called him – hit upon a simple way to do it: Hertting removed the superior cervical ganglion of the cat from one side of the cat, but not the other. Within 2 days the nerves running to the nictitating membrane and the musculus dilator pupillae degenerated. The narrow pupilla of this eye showed this. Tritiated noradrenaline was injected and the amount of tritiated noradrenaline was measured in the nictitating membrane. It had decreased from 45 ng in the intact side to 3.2 ng in the denervated side because no uptake was then possible [Hertting et al., 1961; Pichler et al., 1968; see Kanigel, 1986].

The exact knowledge of the physiology of all these events led to the possibility of interventions by different pharmacological agents which are described briefly in the following.

(4) Reserpine, one of the many alkaloids in *Rauwolfia serpentina*, by blocking vesicular uptake of amines, first releases and then depletes for a long time serotonin and catecholamines from storage vesicles. It was transiently used in the treatment of hypertension [Pletscher et al., 1955; Holzbauer and Vogt, 1956].

(5) Compounds, like ephedrine and amphetamine, displace catecholamines, thus enhancing their release from nerve terminals. Amphetamine releases not only noradrenaline but also dopamine and serotonin in the central nervous system. Repeated administration leads to marked tolerance by depletion of the releasable stores of amines.

The re-uptake of catecholamines into the nerve terminals is inhibited by cocaine. Therefore more molecules are able to activate the postsynaptic receptors. This explains the 'potentiation' of the effect of adrenaline, first described by Fröhlich and Loewi [1910] and explained by Hertting et al. [1961]. This mode of action is also the essential therapeutic effect of tricyclic antidepressant drugs like desipramine.

(6) α-Methyldopa is taken up by noradrenergic neurons and accumulates in synaptic ves-

icles: by decarboxylation and hydroxylation a 'false transmitter' is formed, which is released instead of noradrenaline. It is less active than noradrenaline in causing vasoconstriction. But it is more active on presynaptic α_2-receptors, and greatly increases the autoinhibitory feedback mechanism, thus resulting in a diminished noradrenaline release. Its use is, due to certain side effects, limited to hypertension in late pregnancy.

Clonidine has a peculiar story. It was observed that it contracts the musculi erectores pilorum, which facilitates the effect of an electric shaver. It was planned to market it as a 'preshave', but men found an 'aftershave' would already be enough chemistry on their faces. As clonidine easily penetrates the mucosa and has a long duration of action, the idea to use it for nasal decongestion arose, and a fairly concentrated solution for use as a spray was prepared. A bottle of this solution stood on the table of a pharmacologist who suffered from acute rhinitis. He took the opportunity and put several drops into his nose; he used much more than the dose to be applied by the spray. His colleagues found him fast asleep, suggesting its use in the treatment of insomnia. This made a careful investigation of the circulatory effects necessary, which was done by Kobinger and Pichler [for a review, see 1990] in dogs. They found, after an initial rise, a long-lasting fall in blood pressure. Clonidine, now in the shadow of β-adrenoreceptor blockers, is still an alternative in the treatment of hypertension.

β-Adrenoceptor antagonists were, from a practical therapeutic point, the most important finding in the field of pharmacological influences on the adrenergic system. Patients with hypertension show a gradual fall in arterial blood pressure. The complex action includes a reduction of the cardiac output, reduced renin release from the kidneys, and central sedative actions, based on limited sympathetic activity. Reflex vasoconstriction is preserved and postural or exercise-induced hypotension is much less troublesome than with many other antihypertensive drugs. Although the study of the adrenergic receptor blockade started with Dale's [1906] discovery of the α-adrenergic inhibition by ergotoxine, the practical therapeutic advantage was finally in favour of the β-adrenergic blockade.

Monoamine oxidase (MAO) is an enzyme, widely distributed in all kinds of tissues, which transforms phenylethylamines and serotonin to acidic compounds to be excreted by the urine. MAO-A has a substrate preference for serotonin, MAO-B for phenylethylamine derivatives. MAO inhibitors cause a rapid and sustained increase in the content of serotonin $>$ noradrenaline $>$ dopamine in the brain. They increase the releasable proportion of amines, and thus enhance their effects.

A clinical use of MAO inhibitors was discovered by coincident. In the clinical trial of the tuberculostatic drug iproniazid, the mood of a treated depressive patient was much better than his physical state [Kobinger and Pichler, 1990]. Iproniazid was found to be an inhibitor of MAO. In normal human subjects, the inhibitors cause an immediate increase in motor activity; euphoria and excitement develop in the course of a few days. MAO inhibitors were first introduced as antidepressants, but were largely superseded by tricyclic and other antidepressants [Rang et al., 2003].

Some Peculiarities of Phenylethylamine-Related Substances

Adrenaline was the first hormone to be isolated in crystalline form, acting exclusively in the autonomic and central nervous system. Curiously, adrenaline is also found in the secretions of glands of toads and in the parotid gland of a Jamaican toad *(Bufo marinus)* along with substances resembling cardiac glycosides. A single gland contains more than 4 times the amount of hormone present in a pair of human suprarenal glands [Goodman and Gilman, 1941].

Tyramine is produced from tyrosine by microbial decarboxylation [Dale and Dixon, 1909].

It is found in all kinds of food, and in high amounts in certain types of cheese. During digestion it is already destroyed by the high content of MAO in the intestinal mucosa. Under inhibition of this enzyme by a MAO inhibitor, tyramine is absorbed and causes a so-called 'cheese reaction', consisting of a rise in blood pressure, throbbing headache and even intracranial haemorrhage.

Serotonin which is formed in high amounts in certain plants, such as walnuts [Lembeck and Skofitsch, 1984], is also cleaved in the gut by MAO, but no reports have been made about side effects in patients eating walnuts when on MAO inhibitors.

Ephedrine (β-hydroxy-N-methyl-phenylethylamine) is an alkaloid obtained from plants of the *Ephedra* species; preparations from this plant have been used in China for at least 5,000 years under the name ma-huang, meaning 'yellow horse', which refers to the colour and shape of the flower of *Ephedra sinica.* The main uses of *Ephedra* preparations in Chinese medicine were in the treatment of disorders of the respiratory tract. Ephedrine was introduced into western medicine in 1923, and was the first of the sympathomimetics to be used; in the meantime it has been superseded by more specifically acting compounds.

Cathin (norpseudoephedrine) is contained in the plant khat *(Catha edulis),* a native of Northeast Africa. Khat has a mild psychostimulant action, and lessens the feelings of fatigue and hunger. These effects can be regarded as useful and appropriate under the prevailing conditions of life of the vast majority of users who regularly chew its fresh twigs, leaves and buds. There is a considerable traffic across the Red Sea and down the east coast of Africa.

Mescaline (trimethoxyphenethylamine) was isolated in 1896. In the early 1950s, Aldous Huxley induced a wide interest in the drug. In his excellent novel, he described his experience with mescaline in *Doors of Perception.* There was no loss of intellect or memory, but intensified visual impressions, and a distorted perception of space and time. The motivation to act was replaced by introspection. A mystic identification of self with the surroundings occurred.

Octopamine has been isolated from the posterior salivary glands of *Octopus vulgaris* by Erspamer [1952]. It acts partly directly and partly indirectly on α-receptors and has no clinical value. But octopaminergic neurotransmission exists in insects. The initiation of light emission in fireflies is mediated solely by octopaminergic neurons to the light organ [Robertson and Carlson, 1976]. Cocaine's pesticidal effect is based on an inhibition of octopamine re-uptake, resulting in the potentiation of insect octopaminergic neurotransmission [Nathanson et al., 1993].

Ecstasy (methylenedioxymethamphetamine), a synthetic compound, has actions comparable to amphetamine or methamphetamine. In contrast to the indirectly acting α-adrenergic agonist amphetamine, ecstasy has an indirect serotoninergic effect. Frequent abuse exists today and severe intoxications are not rare.

In conclusion, it can be stated that there are quite a number of therapeutic applications of these substances but also of ways of abuse, which have emerged from the enormous amount of experimental work done by three generations of scientists.

References

Alquist RP: A study of adrenotropic receptors. Am J Physiol 1954;153:586–600.

Aronson JK: 'Where name and image meet' – the argument for 'adrenaline'. BMJ 2000;320:506–509.

Axelrod J: O-Methylation of epinephrine and other catechols in vitro and in vivo. Science 1957;126:400–401.

Axelrod J: Purification and properties of phenylethanolamine-N-methyl transferase. J Biol Chem 1962;237:1657–1660.

Baqu ZM: La pharmacologie du système nerveux autonome, et particulièrement du sympathique d'après la théorie neurohumorale. Ann Physiol Physiochim 1934;10:467–528.

Barger G, Dale HH: Chemical structure and sympathomimetic action of amines. J Physiol 1910;41:19–59.

Bertler A, Rosengren E: Occurrence and distribution of dopamine in brain and other tissue. Experientia 1957;15:10–11.

Blaschko H: Amine oxidase and amine metabolism. Pharmacol Rev 1952;4:415–458.

Bülbring E, Burn JH: Liberation of noradrenaline from the suprarenal gland. Br J Pharmacol 1949;4:202–208.

Cannon WB: Die Notfallsfunktion des sympatho-adrenalen Systems. Ergebn Physiol 1928;27:380–406.

Carlsson A: The occurrence, distribution and physiological role of catecholamines in the nervous system. Pharmacol Rev 1959;11:490–493.

Carlsson A, Falck B, Hillarp NA: Cellular localization of brain monoamines. Acta Physiol Scand 1962;56(suppl 196):1–27.

Cotzias GC, Van Woert, Schiffer LM: Aromatic amino acids and modification of parkinsonism. N Engl J Med 1967;276:374–379.

Dahlström A, Fuxe K: The autonomic nervous system and the histochemical fluorescence method for the microscopical localization of catecholamines and serotonin. Brain Res Bull 1999;50:265–367.

Dale HH: On some physiological actions of ergot. J Physiol 1906;34:163–206.

Dale HH: The action of extracts from the pituitary body. Biochem J 1909;4:427–447.

Dale HH: On the action of ergotoxine. J Physiol 1913;46:291–300.

Dale HH: Medical research as an aim in life. Edinburgh Med J 1949;56:273.

Dale HH, Dixon WE: The action of pressor amines by putrefaction. J Physiol 1909;39:25–44.

Ehringer RH, Hornykiewicz O: Verteilung von Noradrenalin und Dopamin im Gehirn des Menschen und ihr Verhalten bei Erkrankungen des extra-pyramidalen Systems. Klin Wochenschr 1960;38:1236–1239.

Erspamer V: Identification of octopamine as l-*p*-hydroxyphenylethanolamine. Nature 1952;169:375–376.

Falck B, Hillarp NA, Thieme G, Torp A: Fluorescence of catecholamines and related compounds condensed with formaldehyde. Histochem Cytochem 1962;10:348–354.

Fröhlich A, Loewi O: Über eine Steigerung der Adrenalinempfindlichkeit durch Cocain. Arch Exp Pathol Pharmakol 1910;62:159–169.

Gaddum JH: The action of adrenaline and ergotamine on the uterus of the rabbit. J Physiol 1926;61:141–150.

Gaddum JH, Lembeck F: The assay of substances from the adrenal medulla. Br J Pharmacol 1949;4:401–408.

Gaddum JH, Peart WS, Vogt M: The estimation of adrenaline and allied substances in blood. J Physiol 1949;108:467–481.

Gaddum JH, Schild H: A sensitive physical test for adrenaline. J Physiol 1934;80:9P–10P.

Goldenberg M, Faber M, Alston EJ, Chargaff EC: Evidence for the occurrence of nor-epinephrine in the adrenal medulla. Science 1949;109:534.

Goodman L, Gilman A: The Pharmacological Basis of Therapeutics. New York, Macmillan, 1941, p 396.

Hertting G, Axelrod J, Whitby LG: Effect of drugs on the uptake and metabolism of H^3-norepinephrine. J Pharmacol Exp Ther 1961;134:146–153.

Hillarp AN: Structure and function of the autonomic ground plexus. Acta Physiol Scand 1959;46(suppl 157):1–38.

Holtz P, Credner K, Kroneberg G: Über das sympathomimetische pressorische Prinzip des Harnes ('Urosympathin'). Arch Exp Pathol Pharmakol 1944/1947;204:228–243.

Holtz P, Schümann HJ: Ein neues Hormon des Nebennierenmarks. Naturwissenschaften 1948;35:159.

Holzbauer M, Vogt M: Depression by reserpine of the noradrenaline concentration in the hypothalamus of the cat. J Neurochem 1956;1:8–11.

Hornykiewicz O: From dopamine to Parkinson's disease: a personal research record; in Samson F, Adelman G (eds): The Neurosciences: Paths of Discovery. Boston, Birkhäuser, 1992.

Hornykiewicz O: Levodopa in the 1960s: starting point Vienna; in Poewe W, Lees AJ (eds): 20 Years of Madopar – New Avenues. Basel, Editiones Roche, 1994, pp 11–27.

Iversen LL: Role of transmitter uptake mechanisms in synaptic neurotransmission. Br J Pharmacol 1971;41:571–591.

Kanigel R: Apprentice of Genius. New York, Macmillan, 1986.

Kapeller-Adler R: Amine Oxidases and Methods for Their Study. New York, Wiley, 1970.

Kobinger W, Pichler L: Centrally acting drugs (clonidine, methyldopa, guanfacine). Handb Exp Pharmacol. Berlin, Springer, 1990, vol 93, pp 227–262.

Konzett H: Neue broncholytische hochwirksame Körper der Adrenalinreihe. Naunyn Schmiedebergs Arch Pharmacol 1940a;197:27–40.

Konzett H: Zur Pharmakologie neuer adrenalinverwandter Körper. Naunyn Schmiedebergs Arch Pharmacol 1940b;197:41–56.

Konzett H: On the discovery of isoprenaline. Trends Pharmacol 1981;2:47–49.

Konzett H, Rössler R: Versuchsanordnung zu Untersuchungen an der Bronchialmuskulatur. Naunyn Schmiedebergs Arch Pharmacol 1940;195:71–74.

Lembeck F, Skofitsch G: Distribution of serotonin in *Juglans regia* seeds during ontogenetic development and germination. Z Pflanzenphysiol 1984;114:349–353.

Levi-Montalcini R: In Praise of Imperfection. My Life and My Work. New York, Basic Books, 1988.

Levi-Montalcini R, Angeletti PV: Nerve growth factor. Physiol Rev 1968;48:535–569.

Levi-Montalcini R, Hamburger V: Selective growth stimulating effects of mouse sarcoma on the sensory and sympathetic nervous system of the chick embryo. J Exp Zool 1951;116:321–361.

Levi-Montalcini R, Skaper SD: Nerve growth factor: from neuropeptides to neurokine. Trends Neurosci 1996;19:514–519.

Loewi O: Über die humorale Übertragbarkeit der Herznervenwirkung. I. Mitteilung. Pflügers Arch Ges Physiol 1921;189:239–242.

Loewi O: Über die humorale Übertragbarkeit der Herznervenwirkung. IV. Mitteilung. Pflügers Arch Ges Physiol 1924;203:361–367.

Loewi O: Quantitative und qualitative Untersuchungen über den Sympathikusstoff. Pflügers Arch Ges Physiol 1936; 237:504–514.
Loewi O, Navratil E: Über die humorale Übertragbarkeit der Herznervenwirkung. X. Mitteilung. Über das Schicksal des Vagusstoffs. Pflügers Arch Ges Physiol 1927;214: 689–696.
Montague KR: Catechol compounds in rat tissues and in brains of different animals. Nature 1957;180:244–245.
Nathanson JA, Hunnicut EJ, Kantham L, Scavone C: Cocaine as a naturally occurring insecticide. Proc Natl Acad Sci USA 1993;90:9445–9648.
Oliver G, Schäfer EA: The physiological effects of extracts of the suprarenal capsules. J Physiol 1895a;18:230–276.
Oliver G, Schäfer EA: On the physiological action of extracts of pituitary body and certain other glandular organs. J Physiol 1895b;18:277–279.
Otten U, Goedert R, Mayer N, Lembeck F: Requirement of nerve growth factor for development of substance P-containing sensory neurones. Nature 1980;287: 158–159.
Pichler H, Suko J, Hertting G: Action of drugs upon the formation of tritiated H^3-noradrenaline in the rat. Biochem Pharmacol 1968;17:1329–1337.
Pletscher A, Shore PA, Brodie B: Serotonin release as a possible mechanism of reserpine action. Science 1955;122: 374–375.
Rang HP, Dale MM, Ritter JM, Moore PK: Pharmacology, ed 5. Edinburgh, Churchill Livingstone, 1999.
Robertson HA, Carlson AD: Octopamine: presence in the firefly lantern suggests a transmitter role. J Exp Zool 1976; 195:159–164.
Roe DL: The discovery of dopamine's physiological importance. Brain Res Rev 1999;50:5–6.
Sourkes TL, Gauthier S: Levodopa and dopamine antagonists in treatment of Parkinson's disease; in Parnham MJ, Bruinvels J (eds): Discoveries in Pharmacology. Amsterdam, Elsevier, 1983, vol 1, pp 249–267.
Starke K, Göthert M, Kilbinger H: Modulation of transmitter release by presynaptic autoreceptors. Physiol Rev 1989; 69:864–989.
Sutherland EW, Cori CF: Effect of hyperglycaemic factor and epinephrine on liver phosphorylase. J Biol Chem 1951; 188:531–543.
Sutherland EW, Rall TW: The relation of adenosine 3',5'-phosphate and phosphorylase to the action of catecholamines and other hormones. Pharmacol Rev 1960;12: 265–299.
Udenfriend S: Fluorescence Assays in Biology and Medicine. New York, Academic Press, 1962.
Vogt M: The concentration of sympathin in different parts of the central nervous system under normal conditions and after the administration of drugs. J Physiol 1954; 123:451–481.
von Euler US: A specific sympathomimetic ergone in adrenergic nerve fibres (sympathin) and its relation to adrenaline and nor-adrenaline. Acta Physiol Scand 1946;12: 73–97.
Weil-Malherbe H, Bone AD: Intracellular distribution of catecholamines in the brain. Nature 1957;180:1050–1051.

Further Reading (Reviews Published in the Handbook of Experimental Pharmacology)

Blaschko H, Muscholl E: Catecholamines. Handb Exp Pharmacol. Berlin, Springer, 1972, vol 33.
Szekeres L: Adrenergic activators and inhibitors. Handb Exp Pharmacol. Berlin, Springer, 1980, vol 54/I.
Szekeres L: Adrenergic activators and inhibitors. Handb Exp Pharmacol. Berlin, Springer, 1981, vol 54/II.
The IUPHAR Compendium on Receptor Characterization and Classification. London, IUPHAR Media, 2000.
Trendelenburg P: Adrenalin und adrenalinverwandte Substanzen. Handb Exp Pharmacol. Berlin, Springer, 1924, vol 2, part 2.
Trendelenburg U, Weiner N: Catecholamines. Handb Exp Pharmacol. Berlin, Springer, 1988, vol 90/I.
Trendelenburg U, Weiner N: Catecholamines. Handb Exp Pharmacol. Berlin, Springer, 1989, vol 90/II.

Enteramine and Serotonin: Two Names and Three Functions of 5-Hydroxytryptamine

At present, the general interest in serotonin centres around its neuronal functions, especially those implicated in mood regulation. Serotonin-agonistic and -antagonistic drugs, and other drugs which inhibit, more or less specifically, the re-uptake of serotonin into neurons, or induce its release or inhibit its storage, are known. The broad spectrum of therapeutic applications of these drugs includes their use in migraine, in mild depression and other types of anxiety, and as an anti-emetic or stimulator of intestinal peristalsis.

The discovery of serotonin emerged from three independent different lines of research: (1) the histology of enterochromaffin cells in the intestinal mucosa which were first described by Kultschitzky [1897], (2) the histopathological and clinical investigations of carcinoid tumours originating from enterochromaffin cells, and (3) the investigation of vasoconstrictor agents in blood.

Only the essential steps of these investigations are described in the following [for more details, see Creutzfeld and Stockmann, 1987; Erspamer, 1966; Garattini and Valzelli, 1965; Kaehler and Heilmeyer, 1961; Page, 1968].

'Enteramine'

Maffo Vialli, director of the Institute of Comparative Anatomy in Pavia, Italy, had a profound interest in the enterochromaffin cells which were first described by Kultschitzky in 1897. He found the fundus of the stomach of rabbits to be especially rich in these cells. He joined the pharmacologist Vittorio Erspamer to investigate the endocrine functions of this 'enterochromaffin system' [Vialli and Erspamer, 1933]. This resulted in the isolation of this 'hormone'. Erspamer prepared extracts from the rabbit stomach and performed a pharmacological screening on particularly sensitive isolated organs (rat ileum and uterus) and on the rabbit's blood pressure. Erspamer [1940] could differentiate the action of this compound from acetylcholine, histamine, adenosine compounds, tyramine, adrenaline and substance P. It was present only in the gastrointestinal mucosa and in the spleen, but not in any other organs. He provided a basic pharmacological characterization of the compound. Based on simple histochemical tests and chemical properties, like diazo reaction and fluorescence, it could be assumed that the active compound consisted of a phenolic structure linked to an amino group. It was, therefore, named 'enteramine'. The work found hardly any recognition during the war and in the following years, as it was published in the German *Archiv für Experimentelle Pathologie und Pharmakologie*, not available to Anglo-American scientists.

Vittorio Erspamer (1909–1999).

Vasoconstrictor Agents in Blood

Irvine H. Page, at the Cleveland Clinical Center, had the profound idea that hypertension might be caused by a vasoconstrictor agent in blood [Page, 1968; Page and Bumpus, 1974]. There were two possibilities, either angiotensin, formed by renin, or the so-called 'Spätgift', formed during blood clotting. The first line led much later to the most efficient therapy of hypertension based on the use of angiotensin-converting enzyme (ACE) inhibitors.

Renin-Angiotensin

Tigerstedt and Bergmann [1898] had observed that renal ischaemia induced hypertension in dogs. The substance formed in the ischaemic kidney was called renin. Helmer and Page [1939; for a review, see Page and Bumpus, 1974] found that purified preparations of renin were without vasoconstrictor activity in the vessels of rabbit's ear or dog's tail, when perfused with Ringer's solution. The addition of plasma, blood or serum globulin to the purified renin preparation restored its vasoconstrictor property. Page [1968] mentioned later that he switched in his bioassay of the plasma vasoconstrictor from the dog's tail to the rabbit's ear – it was during the Great Depression – for economic reason, because there are two ears rather than one dog's tail used by his colleagues. He used the same preparation later, in the course of the isolation of serotonin.

Today, we know that renin is a proteolytic enzyme that is secreted into the circulation by the granular juxtaglomerular cells of the kidney, acting on a plasma globulin, splitting off a decapeptide, angiotensin I, which is converted by the ACE to the octapeptide, angiotensin II, which is a potent vasoconstrictor. The successful therapeutic use of ACE inhibitors shows that angiotensin II is an essential factor in human hypertension.

'Spätgift'

When freshly collected blood is kept constantly stirred or whipped, the fibrin gradually collects on the surface. In this way in a short period of time all the fibrinogen becomes converted into fibrin, which is removed as it is formed. The fluid which remains is composed of serum together with the red and white cells and resembles blood in appearance and consistency, but of course is incapable of clotting. Thrombocytes are aggregated and destroyed during this procedure. A vasoconstrictor agent, called 'Spätgift', was known to exist in defibrinated blood; it was removed during perfusion through lungs. This was the reason why Starling set up the heart-lung preparation in which defibrinated blood could be used at the time when heparin was still unknown.

From 'Spätgift' to Serotonin

Later on, Page formed a team with Rapport and Green to elucidate the structure of what had been called 'Spätgift'. It resulted in the isolation, crystallization and characterization of the structure of this agent. It was a lucky choice that they started the isolation with 2,000 litres of beef blood, because it was shown later that beef blood contains the highest amount of this vasoconstrictor of all species. They gave the isolated compound the charming name serotonin. Erspamer and Asero [1952] identified serotonin as enteramine. Page, who was unaware of Erspamer's earlier work, commented: 'By any standards, Erspamer must be recognized as one of the most productive and gifted investigators in the field.'

The synthesis of serotonin opened a wide experimental approach during the following three decades. Only the main features can be mentioned here.

About 90% of the total amount of serotonin in the body is present in the enterochromaffin cells interspread with mucosa cells of the stomach and small intestine. These cells are derived from the neural crest, similar to those of the adrenal medulla. Some serotonin also occurs in the myenteric plexus where it functions as an excitatory transmitter. The main intestinal function of serotonin released from enterochromaffin cells in their surrounding is an increased water transfer from the blood into the intestinal lumen, existing under normal conditions and under the influence of cholera toxin [Lembeck et al., 1964; Sewing et al., 1965; for a review, see Beubler 1995].

Platelets accumulate serotonin by an active transport from the plasma [Born et al., 1958; Born and Gillson, 1959; Born, 2005]. When platelets aggregate they release serotonin, as is also the case during blood clotting; the serum contains all the serotonin, which explains the occurrence of 'Spätgift' in serum. Platelets are continuously destroyed in the spleen; this explains the high content of enteramine in the spleen, as found by Erspamer [1940]. Mast cell granula, only in rats and mice, contain more serotonin than histamine; both are released by any kind of degranulation of mast cells, like in inflammation or acute allergic reaction. Under these conditions serotonin contributes to plasma extravasation and formation of oedema in the tissue.

The biosynthesis of serotonin and its degradation closely follow that of noradrenaline. Serotonin is transported into serotonin-containing neurons by a specific transport system. Seven types of serotonin receptors (5-HT_{1-7}) with further subtypes (A–D), all of which but one are G protein-coupled receptors, have been found so far.

Serotonin in the Central Nervous System

The interest in functions of serotonin in the central nervous system arose from the following discoveries:

(1) Amin et al. [1954] investigated the distribution of serotonin and substance P in the central nervous system. Fuxe et al. [1967] later mapped the serotonin-containing neurons in the central nervous system by an improved histochemical method.

(2) Pletscher et al. [1955] found the depletion of serotonin as the essential mechanism of reserpine's action, in parallel with the depletion of noradrenaline.

(3) Gaddum [1958] showed that actions of serotonin are antagonized by the hallucinogen LSD.

(4) Andén et al. [1964] finally defined serotonin as a transmitter in the central nervous system.

(5) Only recently have animal experiments documented how serotonin is implicated in mood regulation; serotonin$_{1A}$ receptors act during the development phase to establish normal anxiety-like behaviour in the adult.

Expression of serotonin$_{1A}$ receptors primarily in the hippocampus and the cortex, but not in the raphe nuclei, was shown to be sufficient to improve the increased anxiety-like behaviour of knockout mice [Gross et al., 2002]. These findings imply the possibility that certain changes in the brain during an early stage of development might be the basis for the development of anxiety and depression many years later, and might influence the further therapeutic development. This means that serotonin, apart from its action by the release from enterochromaffin cells, has a different physiological function as a neurotransmitter in the central nervous system.

Serotoninergic pathways in the brain were found to emerge from the raphe nuclei and to differ from the pathways of acetylcholine, noradrenaline and dopamine. Serotonin receptors in the brain could be localized [Zifa and Fillion, 1992]. These localizations and electrophysiological studies made it possible to derive physiological and behavioural aspects, such as the involvement of serotonin in behavioural changes and hallucinations, sleep and wakefulness, mood and control of sensory transmission. Therapeutic applications have emerged from these investigations in recent years [see Sjoerdsma and Palfreyman, 1990]:

(1) The specific agonist sumatripan is used in migraine therapy.

(2) Serotonin antagonists, like ondansetron, are used as anti-emetics in patients during chemotherapy.

(3) Selective serotonin re-uptake inhibitors, like fluoxetine, are used as antidepressants and in states of anxiety.

Carcinoids

The name carcinoid was created by Oberndorfer [1907] and indicates that the tumour differs in many respects from carcinomas. Carcinoids emerge from enterochromaffin cells which are distributed in the mucosa of the gastrointestinal tract of all vertebrates and many other species (except Cyclostomata and Teleostia). The growth of the tumours is very slow and it might take up to 25 years until metastases in the liver and other organs develop. The shape of enterochromaffin cells with granula at their base indicated that they might release a substance into their surroundings ('paracrine') or into the blood ('endocrine'). The Austrian pathologist Feyrter [1934] concluded that these cells represent a diffusely distributed endocrine organ. Feyrter [1938] tested the extract of a carcinoid on the blood pressure of the cat and found a rise in blood pressure similar to that evoked by adrenaline, an action later confirmed by Lembeck [1954].

Cassidy (1931, 1932) described a 'phenomenal flushing' in the face of a carcinoid patient with metastases; the event was so outstanding that a colour portrait of an artist was included in his publication. Isler and Hedinger described the massive diarrhoea in carcinoid patients. The full 'triad' of the clinical syndrome in carcinoid patient was described by Waldenström et al. [1956]. It consists of (1) flushes in 75%, (2) diarrhoea in 84% of carcinoid patients and (3) endocard fibrosis, predominantly in the right heart, in 45%. Elevated urinary excretion of 5-hydroxyindoleacetic acid, the end product of serotonin metabolism, was found increased in 84% of the patients; the test is used to confirm the diagnosis. The symptoms of the carcinoid syndrome can be attributed to serotonin. (1) Intracutaneous injection of serotonin causes vasodilatation and oedema. (2) Serotonin, in a dose which does not yet contract smooth muscles of the rat intestine, increases the water secretion into the gut. The diarrhoea can, therefore, not be attributed to a serotonin-mediated increase in intestinal motility (as explained in current textbooks), but to the filling of the intestinal lumen with fluid, by which the peristaltic reflex is evoked. (3) The endocard fibrosis in the right heart is based on the high concentration of serotonin in the venous outflow from the liver metastases which reaches the right heart, whereas the left heart, after the passage through the lungs where serotonin is destroyed by monoaminooxidase, is reached only by much lower concentrations. In half of the carcinoids, tachykinin peptides were found in the tumours and were also co-secreted with serotonin. This indicates that carcinoids originate from a very early stage of the development of the enterochromaffin cells: secreted tachykinins might also contribute to the clinical symptoms of carcinoids [Norheim, 1986].

My personal contact with serotonin was a lucky event (F.L.). Loewi and Hellauer [1938] had shown that the dorsal roots of the spinal cord, which carry afferent fibres only, contain no acetylcholine. This was in contrast to the ventral roots carrying only efferent motor fibres containing acetylcholine. I found that the extracts of dorsal roots were active to produce a contraction of the isolated guinea pig ileum and could, with the use of antagonists, exclude acetylcholine and histamine and assumed that only substance P could be the responsible agent [Lembeck, 1953]. I showed these results to Sir John Gaddum, when he came for a short visit to Graz. He asked, 'Could you exclude serotonin?' and my reply was the question, 'What is serotonin?'. He explained to me the work of Erspamer and the recent synthesis of serotonin. Immediately, I wrote a very polite letter to Dr. Weisblat at the Upjohn Company who was the first to have synthesized serotonin [Speeter et al., 1951]. He promptly sent me 10 mg, and I was able to exclude the presence of serotonin in dorsal root extracts.

Serotonin was a gift which arrived at the right time for me. My colleague Max Ratzenhofer, pathologist and former coworker of Feyrter, had already told me long stories about the pathology of carcinoids and now he presented me with the liver metastasis of a carcinoid just obtained from a postmortem dissection. I think I was at that time the only person on the European continent to possess the treasure of a few milligrams of serotonin. It was, therefore, easy to measure its high concentration in the carcinoid metastasis [Lembeck, 1953]. I used my publication to apply for the academic title of 'Dozent', which also meant that I had to give a lecture. To my surprise, the lecture was arranged in a clinical lecture theatre, and was given the terrible title 'About Serotony' (comparable to dysentery), because the Dean thought I would be talking about a new type of disease!

Serotonin, Somewhere Else

Serotonin is widely distributed in different organs of many species; the highest amount was found in the skin of *Bufo alvaris*, a toad from Arizona [Page, 1968].

Exclusively in the pineal gland, melatonin is formed from serotonin by acetylation and O-methylation of serotonin. Melatonin release into the circulation is high at night, and low by day. The circadian rhythm of the release is controlled by neuronal input from the retina [Brzezinski, 1997].

Wasp venom *(Vespula vulgaris)* contains Thr^6-bradykinin, histamine, and serotonin [Jaques and Schachter, 1954]; the pain is caused by the peptide, the inflammation mainly by serotonin [Griesbacher et al., 1998].

It should be mentioned that serotonin occurs in quite a different location in nature. It has been found in the vesicles of nettle stings along with acetylcholine and histamine. If you come into contact with a leaf, the sharp tips of such vesicles break and you receive an intracutaneous injection of the three ingredients, acetylcholine, serotonin and histamine – with the effect known to every child [Collier and Chesher, 1956]. Peripheral terminals of thin afferent nerve fibres are highly sensitive to heat and to many chemical compounds, such as bradykinin, acetylcholine, serotonin and histamine [Lembeck and Juan, 1974].

In an insane asylum in Frankfurt, Kirberger [1962] measured the final metabolic product of serotonin excreted in the urine, hydroxyindoleacetic acid. He collected samples of urine of all available patients to find a possible relation to their clinical condition. In one patient he discovered waves of a highly increased excretion of hydroxyindoleacetic acid on the days following his mother's visits. She had brought walnuts which he liked so much that he consumed them within hours. Walnuts contain remarkable quantities of serotonin, which is also found in all parts of the walnut tree [Lembeck and Skofitsch, 1983]. Oral consumption has no systemic effect since it is already destroyed by the monoaminooxidase of the intestinal epithelium and transformed into inactive metabolites.

Derivatives and structural analogues of serotonin can produce hallucinogenic effects. Psilocybin, 4-hydroxytryptamine phosphate, has been isolated from a Mexican mushroom. It is the phosphate ester of 4-hydroxytryptamine. Since prehistoric times, it has been given by shamans, usually women, to suitable test persons in ritual ceremonies at night; the ecstatic reactions of these people under the influence of the drug were interpreted for the audience by the shaman. In the meantime psilocybin was found in many of the 150 species of *Psilocybe* mushrooms around the world [Weil and Rosen, 1998], including *Psilocybe semilanceata* in Europe [Schmid and Helfer, 1995].

Dimethyltryptamine and 5-methoxy-dimethyltryptamine were found in the resin of a huge jungle tree in the Amazonian forest, from which a dried extract was prepared – the obtained powder, called yupo, was sniffed. The hallucinogenic effect starts immediately and lasts for 15 min; therefore, it is called a businessman's trip. A toad, living the Sonoran Desert of Arizona, produces large amounts of 5-methoxy-dimethyltryptamine secreted from cutaneous glands [Weil and Rosen, 1998].

References

Amin AH, Crawford TBB, Gaddum JH: The distribution of substance P and 5-hydroxytryptamine in the central nervous system of the dog. J Physiol 1954;126:596.

Andén NE, Carlsson A, Hillarp NA, Magnusson T: 5-Hydroxytryptamine release by nerve stimulation of the spinal cord. Life Sci 1964;3:473–478.

Beubler E: Serotonin as an intestinal secretagogue; in Gagniella TS, Galligan JJ (eds): Serotonin and Gastrointestinal Function. Boca Raton, CRC Press, 1995, pp 85–108.

Born GVR: Light on platelets. J Physiol, in press.

Born GVR, Gillson RE: Studies on the uptake of 5-hydroxytryptamine by blood platelets. J Physiol 1959;146:472–491.

Born GVR, Ingram GIC, Stacey RS: The relationship of 5-hydroxytryptamine and adenosine triphosphate in blood platelets. Br J Pharmacol 1958;13:62–64.

Brzezinski A: Melatonin in humans. N Engl J Med 1997;336: 186–195.

Collier HOJ, Chesher GB: Identification of 5-hydroxytryptamine in the sting of the nettle *(Urtica dioica)*. Br J Pharmacol 1956;11:186–189.

Creutzfeld W, Stockmann F: Carcinoids and carcinoid syndrome. Am J Med 1987;82(suppl 58):4–16.

Erspamer V: 5-Hydroxytryptamine and related indolethylamines. Handb Exp Pharmakol. Berlin, Springer, 1966, vol 19.

Erspamer V: Pharmakologische Studien über Enteramin. Naunyn Schmiedebergs Arch Pharmacol 1940;196:343–407.

Erspamer V, Asero B: Identification of enteramine, specific hormone of the enterochromaffin system, as 5-hydroxytryptamine. Nature 1952;169:800.

Feyrter F: Carcinoid und carcinom. Ergeb Pathol 1934;29: 305.

Feyrter F: Über diffuse endokrine epitheliale Organe. Leipzig, Johann Ambrosius Barth Verlag, 1938.

Fuxe K, Hökfelt T, Ungerstedt U: Morphologic, pharmacologic and functional studies on 5-hydroxytryptamine neurons in the central nervous system of mammals. 5-HT Symposium, New York, 1967.
Gaddum HJ: Drugs which antagonize the actions of 5-hydroxytryptamine on peripheral tissues; in Lewis GP (ed): 5-Hydroxytryptamine. London, Pergamon Press, 1958.
Garattini S, Valzelli L: Serotonin. Amsterdam, Elsevier, 1965.
Griesbacher T, Althuber P, Zenz M, Rainer I, Griengl S, Lembeck F: *Vespula vulgaris* venom: role of kinins and release of 5-hydroxytryptamine from skin mast cells. Eur J Pharmacol 1998;351:95–104.
Gross C, Zhuang X, Stark K, Ramboz S, Oosting R, Kirby L, Santarelli L, Bech S, Hen R: Serotonin$_{1A}$ receptor acts during development to establish normal anxiety-like behaviour in the adult. Nature 2002;416:396–400.
Jaques R, Schachter M: The presence of histamine, 5-hydroxytryptamine and a potent, slow contracting substance in wasp venom. Br J Pharmacol 1954;9:53–58.
Kaehler HJ, Heilmeyer L: Clinical aspects and pathophysiology of carcinoid and carcinoid syndrome with special reference to the pharmacology of 5-hydroxytryptamine. Ergeb Inn Med Kinderheilk 1961;16:292–449.
Kirberger E: Differentialdiagnostische Überlegungen bei vermehrter 5-Hydroxyindolessigsäure-Ausscheidung im Harn. Dtsch Med Wochenschr 1962;87:929–934.
Kultschitzky N: Zur Frage über den Bau des Darmkanals. Arch Mikr Anat 1897;49:7–35.
Lembeck F: Zur Frage der zentralen Übertragung afferenter Impulse. III. Das Vorkommen und die Bedeutung der Substanz P in den dorsalen Wurzeln des Rückenmarks. Naunyn Schmiedebergs Arch Exp Pathol Pharmakol 1953a;219:197–213.
Lembeck F: 5-Hydroxytryptamine in a carcinoid tumour. Nature 1953b;172:910–911.
Lembeck F: Über den Nachweis von 5-Oxytryptamin (Enteramin, Serotonin) in Carcinoidmetastasen. Arch Exp Pathol Pharmakol 1954;221:50–66.
Lembeck F, Juan H: Interaction of prostaglandins and indomethacin with algesic substances. Naunyn Schmiedebergs Arch Pharmacol 1974;285:301–313.
Lembeck F, Sewing KF, Winne D: Der Einfluss von 5-Hydroxytryptamin auf die Resorption von Tritium-Wasser (HTO) aus dem Dünndarm der Ratte. Naunyn Schmiedebergs Arch Pharmacol 1964;247:100–109.
Lembeck F, Skofitsch G: Distribution of serotonin in *Juglans regia* seeds during ontogenetic development and germination. Z Pflanzenkd 1983;114:349–353.
Loewi O, Hellauer H: Über das Acetylcholin in peripheren Nerven. Pflügers Arch Ges Physiol 1938;240:769–775.
Norheim I: Carcinoid tumors: clinical studies and the use of tachykinins as tumor markers. Acta Universitatis Uppsaliensis 1986;40.
Oberndorfer S: Karzinoide Tumoren des Dünndarms. Frankf Z Pathol 1907;1:426.
Page IH: Serotonin. Chicago, Year Book Medical Publishers, 1968.
Page IH, Bumpus FM: Angiotensin. Handb Exp Pharmakol. Berlin, Springer, 1974, vol 37.
Pletscher A, Shore PA, Brodie BB: Serotonin release as a possible mechanism of reserpine action. Science 1955;122: 374–375.
Schmid H, Helfer W: Wissenswertes aus Ökologie, Geschichte und Mythos. Eching bei München, IHW-Verlag, 1995.
Sewing KF, Lembeck F, Winne D: The effect of 5-hydroxytryptamine on the resorption of tritiated water (HTO) from the rat colon in vivo. Naunyn Schmiedebergs Arch Exp Pathol Pharmakol 1965;252:286–290.
Sjoerdsma A, Palfreyman MG: History of serotonin and serotonin disorders. Ann NY Acad Sci 1990;600:1–8.
Speeter ME, Heinzelmann RV, Weisblat DI: The synthesis of the blood serum vasoconstrictor principle serotonin creatinine sulphate. J Am Chem Soc 1951;73:5514–5515.
Tigerstedt R, Bergmann P: Niere und Kreislauf. Skand Arch Physiol 1898;8:223–271.
Vialli M, Erspamer V: Cellule enterocromaffini e cellule basigranulose acidofile nei vertebrati. Z Zellforsch 1933;19: 743–773.
Waldenström J, Pernow P, Silver H: Case of metastasizing carcinoma (argentaffinoma) of unknown origin showing peculiar red flushing and increased amounts of histamine and 5-hydroxytryptamine in blood and urine. Acta Med Scand 1957;156:73–83.
Weil A, Rosen W: From Chocolate to Morphine. Boston, Houghton Mifflin, 1998.
Zifa E, Fillion G: 5-Hydroxytryptamine receptors. Pharmacol Rev 1992;44:401–458.

Histamine: One Substance and Three Functions

β-Aminoethylimidazol is derived from histidine by decarboxylation. It was first synthesized by Windaus and Vogt [1907] and was soon afterwards named histamine. Like serotonin, it shows neuronal and extraneuronal actions.

The role of histamine is reviewed by Schwartz et al. [1995]. Histaminergic pathways have been described with cell bodies in a small region of the hypothalamus. The role of histamine in the central nervous system remains poorly understood and less defined than that of other neurotransmitters. The sedative and anti-emetic side effects of H_1 receptor antagonists are the few existing hints for central actions [Rang et al., 2003].

Stimulation of peripheral cutaneous terminals of afferent fibres causes the 'triple response' (according to Lewis) with local vasodilatation and wheal including increased plasma extravasation; the surrounding flare results from an 'axon reflex' by the local release of peptide mediators [Lembeck, 1983]. The involved nerve stimulation by histamine is not specific as the terminals of C fibres are also stimulated by bradykinin, acetylcholine, serotonin, and ATP [Juan and Lembeck, 1976; Lembeck et al., 1976]. Itching is observed by cutaneous injection of histamine or by touching nettles, from which a mixture of histamine, acetylcholine and serotonin is injected intracutaneously.

Histamine is stored in the granula of mast cells. Release from mast cells is triggered experimentally by a number of basic compounds, or when antigen interacts with cell-fixed IgE, which causes urticaria in human skin.

In 1913 it was known that anaphylactic shock resulted from the immediate reaction between antibody and antigen. Dale wanted to know whether it is based on the circulating fluids or on responsive tissues. It was known at the Wellcome Laboratories which produced several antisera that a guinea pig sensitive to horse serum may be killed in a few minutes by intravenous injection of 0.01 ml of the serum. He had observed that during the anaphylactic reaction in vivo the uterus appeared to become oedematous. Therefore, he used the isolated uterus of guinea pigs to test freshly prepared blood sera, of which horse serum was the most easily available, and compared it with the effect of histamine. He had first isolated and chemically identified histamine from beef intestine [Barger and Dale, 1910], and shown that it contracted the uterus [Dale and Laidlow, 1910]. He came across a uterus, from a specific guinea pig, which contracted rapidly and to a maximal extent in response to the first application of horse serum in a normally quite inactive dilution. It was a guinea pig that had survived an antitoxin test and which was regarded to be 'good enough' for his 'pharmacological experiments'! Dale [1913] described the species selectivity of this reaction, discovered the passive sensitization of smooth muscle and pointed out the similarity to the effect of histamine.

When the first generation of antihistaminic drugs was produced, following the work of Bovet, it was found that they antagonized all effects of histamine except the histamine-induced gastric acid secretion which pointed to two different histamine receptors. Hydrochloric acid in the stomach is produced and released from parietal cells upon stimulation by the hormone gastrin, by neurogenic acetylcholine and by histamine, derived from closely neighbouring mast cells, on H_2 histamine receptors. The discovery of H_2 receptor antagonists by James Black, for which he received the Nobel Prize, has been the most rewarding discovery in the broad field of histamine research.

References

Barger G, Dale HH: β-Imidazolethylenamine, a depressor constituent of intestinal mucosa. J Physiol 1910;41:499–503.

Dale HH: The anaphylactic reaction of plain muscle in the guinea-pig. J Pharmacol Exp Ther 1913;4:167–233.

Dale HH, Laidlaw PP: Further observations on the action of β-imidazolinethylamine. J Physiol 1910;43:182–195.

Juan H, Lembeck F: Release of prostaglandins from the isolated perfused rabbit ear by bradykinin and acetylcholine. Agents Action 1976;6:642–645.

Lembeck F: Sir Thomas Lewis's nocifensor system, histamine and substance P-containing primary afferent nerves; in Bousfield D (ed): Neurotransmitters in Action. Amsterdam, Elsevier Science, 1983, pp 173–179.

Lembeck F, Popper H, Juan H: Release of prostaglandins by bradykinin as an intrinsic mechanism of its algesic effect. Naunyn Schmiedebergs Arch Pharmacol 1976;294:69–73.

Rang HP, Dale MM, Ritter JM, Moore PK: Pharmacology, ed 5. Edinburgh, Churchill Livingstone, 2003.

Schwartz JC, Arrang JM, Garbarg M, Traiffort E: Histamine; in Bloom E, Kupfer DJ (eds): Psychopharmacology: A Forth Generation of Progress. New York, Raven Press, 1995.

Windaus A, Vogt D: Synthese des Imidazolyläthylamins. Ber Dtsch Chem Ges 1907;40:3691–3696.

The Amino Acid Transmitter Family

The central nervous system (CNS) contains uniquely high concentrations of certain amino acids, notably glutamate and γ-aminobutyrate (GABA); these amino acids are extremely potent in their ability to alter neuronal discharge. However, many physiologists were extremely reluctant to accept these simple substances as central neurotransmitters. This reluctance was based in part on conceptual problems of how to discriminate amino acids acting as transmitters from the same compounds acting as precursors for protein synthesis. The ubiquitous distribution of amino acids within the brain also posed problems in relating release to activity of a single neuronal circuit. Other important arguments against amino acids as transmitters were that they produced prompt, powerful, and readily reversible but redundant effects on every neuron tested; the dicarboxylic amino acids glutamate and aspartate produce excitation, and the monocarboxylic ω-amino acids GABA and glycine produce a qualitatively similar inhibition.

Could amino acids with major metabolic roles be at variance with the pattern set by an established agent such as acetylcholine (which actually is also an amino acid derivative!)? Curtis et al. [1959] demonstrated that glutamate and aspartate, iontophoretically applied to the dorsal horn of the spinal cord of cats, were found to be very effective in evoking repetitive discharges in interneurons. This was a 'prehistoric' finding [in the words of Krnjevic, 1970]. During the 1960s (the 'Dark Ages') excitatory amino acids and GABA were regarded as pharmacological curiosities. In the 1970s (the 'Renaissance') glycine was established as an inhibitory transmitter in spinal cord interneurons. In the following 'Baroque Era', the physiological role of glutamate was established, an advance mainly based on the discovery of specific antagonists.

During the following decades basically all the conceptual arguments against amino acids as neuronal messengers have proven to be unjustified and certain amino acids are now very well-established neurotransmitters [Bloom, 1985]. They have synthetic companions, pharmacological supporters, specific targets, and some of them powerful opponents.

The amino acid *L*-glutamate is considered to be the major mediator of excitatory signals in the mammalian CNS and is probably involved in most aspects of normal brain function including cognition, memory and learning [Orrego and Villanueva, 1993]. Glutamate also plays a major role in the development of the CNS, including synapse induction and elimination, and cell migration, differentiation and cell death. Most neurons and even glial cells have glutamate receptors in their plasma membranes.

It was noted by Krebs [1935] already in 1935 that glutamate plays a central metabolic role in the brain and that brain tissue has a very high glutamate uptake activity. In search for the main central excitatory transmitter, the work of Hayashi [1954] gave an early indication of the excitatory potency of glutamate, but also of aspartate, α-ketoglutarate, citrate and phosphate. This work was taken up with great vigor by Curtis et al. [1959, 1960], who must be given credit for opening up the field, and who were able to show that numerous amino acids, several of them endogenous to the brain, are able to excite virtually all central neurons. The very extension of the excitatory action led this group to state 'that this action is non-specific and unrelated to excitatory synaptic transmitter action'. This conclusion, which at the time seemed quite reasonable, was not shared, however, by Krnjevic [1965, 1974]. It was soon suggested that glutamate might have a more selective action by demonstrating very specific extracellular actions of *L*-glutamate in thalamic neurons. Knowledge of the structural and functional characteristics of central excitatory neurons also progressed steadily. Thus, the characterization of the excitatory postsynaptic potentials (EPSPs) gave a solid foundation to the very concept of chemical neurotransmission in the CNS. Synaptic regions were shown to contain unique structures, the synaptic vesicles, for which a role in neurotransmitter storage was at once suggested. The mechanisms of transmitter release also became better defined.

The complex compartmentalization of the glutamate-glutamine metabolism was elucidated soon afterwards. These findings led to the understanding that glutamate uptake is important in controlling the excitatory action of glutamate. Synaptic release of glutamate was demonstrated and it became soon clear that an amino acid with multiple 'ordinary' metabolic roles, such as *L*-glutamate, actually could be the major excitatory neurotransmitter in the mammalian nervous system.

L-Glutamate is the most prominent member of the family, its function related with the weaker-acting aspartate and the synthetic N-methyl-*D*-aspartate (NMDA), 10–1,000 times more potent than *L*-glutamate on the specific NMDA receptor. Notwithstanding the high-affinity binding of glutamate to synaptosomes and a calcium- and stimulation-dependent glutamate release, it was finally the accumulation of glutamate, labelled by the immunogold method, in synaptic terminals that established its role as a transmitter [Weinberg et al., 1987].

Otsuka and Konishi [1974] invented one of the most elegant in situ preparations in neuroscience, the hemisectioned isolated spinal cord of newborn rats. Electrical stimulation of the dorsal roots and extracellular recording from the ventral roots made it possible to study the influence of substances applied to the superfusion system of the spinal cord. They found a 200-times higher potency of the depolarizing action of substance P than that of *L*-glutamate, which made them focus on substance P in the following years.

Glutamate is widely and fairly uniformly distributed in the CNS, and its concentration there is much higher than in other tissues. Glutamate in the CNS comes partly from glutamine which is synthesized in glial cells and taken up by the neurons. Glutamate is stored in synaptic vesicles and released upon stimulation of the nerve by a calcium-dependent exocytosis. Its action is terminated by a carrier-mediated uptake into nerve terminals and glial cells. The glutamic acid decarboxylase (GAD) transforms glutamate to the inhibitory transmitter GABA. The GABA transaminase destroys glutamate and GABA. Glutamate is also cleaved by a transaminase which, in parallel, transforms glyoxalate to the inhibitory transmitter glycine.

Aided by the synthesis of a number of different glutamate and aspartate analogues, different types of glutamate receptors were identified pharmacologically [Roberts, 1974]. When the glutamate receptors were cloned in the early 1990s, three different families of glutamate receptor proteins were identified: one family of glutamate receptors is activated by the glutamate analogue NMDA and these receptors are collectively referred to as NMDA receptors. Another family of receptors is activated by α-amino-3-hydroxy-5-methyl-4-isoxazole propionic acid (AMPA) and by kainate; these receptors are subdivided into AMPA receptors and kainate receptors. The NMDA and AMPA/kainate receptors are ion channels and are collectively referred to as ionotropic glutamate receptors. The third family of glutamate receptors consists of G protein-coupled receptors, the so-called metabotropic receptors.

Functional roles of glutamate receptors are involved in the long-term synaptic plasticity, in excitotoxicity and in the pathogenesis of epilepsy. Glycine has a facilitatory influence on the NMDA receptor. Ketamine, used as an anaesthetic, has a selective blocking effect on them.

During the 1970s and 1980s, a number of groups studied the properties of the glutamate uptake system, the only system to remove glutamate from the extracellular fluid surrounding the receptors [Danbolt, 2001]. These transporters are now referred to as 'excitatory amino acid transporters' ('EAATs') and antibodies against them are an elegant way to localize glutamate neurons.

A coexistence of neuropeptides with so-called 'classical' transmitters has been discovered by Hökfelt [for a review, see Hökfelt et al., 1980]. Primary afferent C fibres contain not only the peptides substance P, CGRP, and somatostatin, but also glutamate as shown by immunohistochemistry [De Biasi and Rustioni, 1988; Todd and Spike, 1993]. A stimulus-induced corelease of substance P and glutamate from their central terminals has been shown by Ueda et al. [1993, 1994]. Circulatory depressor reflexes in the rat, induced by various afferent nervous stimuli, were shown to be inhibited by the NMDA antagonist MK-801, but not by the substance P antagonist SR-140333, indicating the essential role of glutamate in the central transmission of such reflexes, and attributing a modulatory role to substance P [Juránek and Lembeck, 1996, 1997].

GABA was identified as a unique chemical constituent of the brain in 1950, but its potency as a CNS depressant was not immediately recognized. GABA has first been found in comparable amounts in normal and degenerated sciatic nerves by paper-chromatographic separation of amino acids [May and Thillard, 1951]. A non-identified inhibitory factor, called factor I [Florey, 1953, 1967], was claimed to be identical to GABA; this, however, could not be substantiated [Florey and McLennan, 1959].

Ernst Florey first used inhibition of the crayfish (a crustacean) stretch receptors as an assay in his discovery that GABA was likely to be an inhibitory neurotransmitter [Florey and Biedermann, 1960]. GABA mimicked the actions of the stimulation of the inhibitory nerve, and picrotoxin antagonized both the effects of applied GABA and stimulation of the inhibitory nerve.

The initial suggestion that GABA was the major inhibitory transmitter was followed by severe doubts that were caused by the relatively large amounts of GABA found in the mammalian brain, the simple nature of this molecule and experiments that seemed to rule it out. At the end of an international conference on GABA, the statement was made that 'GABA entered the conference as a proud transmitter candidate and left it as a poor metabolite' [Edwards et al., 1999].

Kravitz et al. [1963] found that GABA was the only inhibitory amino acid found exclusively in the inhibitory nerve of the crustacean crayfish, but not in the excitatory motor axons. The inhibitory potency of the extracts of the inhibitory nerve was accounted for by their content of GABA. Release of GABA was then correlated with the frequency of nerve stimulation. Intracellular recordings from the muscle indicated that the inhibitory nerve and GABA produced identical increases in Cl^- conductance in the muscle [Otsuka and Konishi, 1974]. These studies definitively resolved the question of the role of GABA as an inhibitory neurotransmitter and GABA was established as the most active inhibitory substance in the CNS.

GABA selectively occurs in nervous tissue, in very high amounts in the nigrostriatal system (10 μmol/g tissue) and in lower concentrations throughout the grey matter (2–5 μmol/g tissue). It has been estimated that GABA serves as a transmitter in 30% of all synapses in the CNS. GABA mediates the inhibitory actions of local interneurons in the brain and it may also mediate presynaptic inhibition. Many GABAergic inhibitory neurons were localized by cytochemical methods.

GABA acts on two types of receptors, $GABA_A$ and $GABA_B$ receptors. $GABA_A$ receptors are anion channels belonging to the same superfamily as other transmitter-gated ion channels in the CNS. On $GABA_A$ receptors, muscimol, derived from a hallucinogenic mushroom, is a powerful agonist, and bicuculline and picrotoxin are antagonists. $GABA_A$ receptors contain an exceptional number and variety of modulatory sites where ligands modify the functional response to GABA. Benzodiazepines, barbiturates and neurosteroids facilitate the binding of GABA on the receptor and enhance its inhibitory transmitter function. It should be mentioned that barbiturates were discovered a hundred years ago and benzodiazepines resulted from a large-scale screening from 1961 onward, leading to a wide range of analogues with broad therapeutic applications. Only the later research on GABA receptors made it possible to explain their mode of action.

The recent elucidation of the existence of multiple subtypes of $GABA_A$ receptors in the brain should facilitate the development of even more selective drugs [Sieghart and Sperk, 2002].

$GABA_B$ receptors are coupled to G protein-gated Ca^{2+} and K^+ channels. Accordingly presynaptic $GABA_B$ receptors influence neurotransmission by the suppression of neurotransmitter and neuropeptide release. An agonist at $GABA_B$ sites is baclofen.

Glycine was found not to be a particularly potent agent when its inhibitory effects were first evaluated by the iontophoretic technique in the spinal cord. However, Werman et al. [1968] assembled neurochemical and electrophysiological evidence that strongly supports a role for glycine as the inhibitory transmitter between spinal interneurons and motoneurons. Glycine exists in amounts of 5 μmol/g in the grey matter of the spinal cord and induces hyperpolarization in interneurons that is indistinguishable from the inhibitory synaptic response. This inhibitory effect of glycine is quite distinct from its facilitatory role in NMDA receptors.

Glycine binds to postsynaptic glycine receptors, causing gating of an integrated ion channel that increases the chloride conductance of the plasma membrane and hyperpolarization [Betz, 1991]. It has a toxic competitive antagonist, the convulsion-producing alkaloid strychnine which blocks both the synaptic inhibitory response and the response to glycine. Tetanus toxin acts mainly by interfering with the release of glycine from inhibitory interneurons in the spinal cord and causes excessive reflex hyperexcitability and violent muscle spasms.

Weinberg [1999] concluded his brief current review with several questions concerning glutamate which also hold for the other amino acid transmitters. He concluded with the following two sentences:

'Current evidence suggests that postsynaptic receptors may be organized into functional microdomains; what scaffolding and signalling molecules are associated with individual receptor clusters? Thus, the excitement produced by glutamate will continue into the next millennium.'

References

Betz H: Structure and function of inhibitory glycine receptors. Q Rev Biophys 1992;25:381–394.

Bloom FE: Neurohumoral transmission and the central nervous system; in Goodman A, Goodman LS, Rall TW, Murad F (eds): The Pharmacological Basis of Therapeutics, ed 7. New York, Macmillan, 1985, pp 236–259.

Curtis DR, Phillis JW, Watkins JC: Chemical excitation of spinal neurons. Nature 1959;183:611–612.

Curtis DR, Phillis JW, Watkins JC: The chemical excitation of spinal neurones by certain acidic amino acids. J Physiol (Lond) 1960;150:656–682.

Danbolt NC: Glutamate uptake. Prog Neurobiol 2001;65: 1–105.

De Biasi S, Rustioni A: Glutamate and substance P coexist in primary afferent terminals in the superficial laminae of the spinal cord. Proc Natl Acad Sci USA 1988;85: 7820–7824.

Edwards DH, Heitler WJ, Krasne FB: Crustacean studies and the early history of GABA. Trends Neurosci 1999;22: 347.

Florey E: Über einen nervösen Hemmungsfaktor in Gehirn und Rückenmark. Naturwissenschaften 1953;40: 295–296.

Florey E: Neurotransmitters and modulators in the animal kingdom. Fed Proc 1967;26:1164–1178.

Florey E, Biederman MA: Studies on the distribution of factor I and acetylcholine in crustacean peripheral nerve. J Gen Physiol 1960;43:509–522.

Florey E, McLennan H: The effect of factor I and of gamma-aminobutyric acid on smooth muscle preparations. J Physiol 1959;145:66–67.

Hayashi T: Effects of sodium glutamate on the nervous system. Keio J Med 1954;3:183–192.
Hökfelt T, Johansson O, Ljungdahl A, Lundberg JM, Schultzberg M: Peptidergic neurones. Nature 1980;284:515–521.
Juránek I, Lembeck F: Afferent C-fibres release substance P and glutamate. Can J Physiol Pharmacol 1997;75:661–664.
Juránek I, Lembeck F: Evidence for the participation of glutamate in reflexes involving afferent, substance P containing nerve fibres in the rat. Br J Pharmacol 1996;117:71–78.
Kravitz EA, Kuffler SW, Potter DD: Gamma-aminobutyric acid and other blocking compounds in Crustacea. Their relative concentration in separated motor and inhibitory axons. J Neurophysiol 1963;26:739–751.
Krebs HA: Metabolism of amino acids. IV. Synthesis of glutamine from glutamic acid and ammonia, and the enzymatic hydrolysis of glutamine in animal tissue. Biochem J 1935;29:1951–1969.
Krnjevic K: Actions of drugs on single neurones in the cerebral cortex. Br Med Bull 1965;21:10–14.
Krnjevic K: Chemical nature of synaptic transmission in vertebrates. Physiol Rev 1974;54:418–540.
Krnjević K: Glutamate and γ-aminobutyric acid in brain. Nature 1970;228:119–124.
May RM, Thillard MJ: Chromatographie des acides aminés du nerf sain et dégénéré. Bull Soc Chim Biol 1951;33:253–257.
Orrega F, Villanueva S: The chemical nature of the main central excitatory transmitter: a critical appraisal based upon release studies and synaptic vesicle localization. Neuroscience 1993;56:539–555.
Otsuka M, Konishi S: Electrophysiology of mammalian spinal cord in vitro. Nature 1974;252:733–734.
Roberts PJ: Glutamate receptors in the rat central nervous system. Nature 1974;252:399–401.
Sieghart W, Sperk G: Subunit composition, distribution and function of $GABA_A$ receptor subtypes. Curr Top Med Chem 2002;2:795–816.
Todd AJ, Spike RC: The localization of classical transmitters and neuropeptides within neurons in laminae I–III of the mammalian spinal dorsal horn. Prog Neurobiol 1993;41:609–645.
Ueda M, Kuraishi Y, Satoh M: Detection of capsaicin-evoked release of glutamate from spinal dorsal horn slices of rat with on-line monitoring system. Neurosci Lett 1993;155:179–182.
Ueda M, Kuraishi Y, Sugimoto K, Satoh M: Evidence that glutamate is released from capsaicin-sensitive primary afferent fibers in the rat: study with on-line continuous monitoring of glutamate. Neurosci Res 1994;20:231–237.
Weinberg RJ: Glutamate: an excitatory neurotransmitter in the mammalian CNS. Brain Res Bull 1999;50:353–354.
Weinberg RJ, Conti F, Van Eyck S, Perusz P, Rustioni A: Glutamate-like immunoreactivity in rat substantia gelatinosa; in Hicks TP, Lodge D, McLennan H (eds): Excitatory Amino Acid Transmission. New York, Liss, 1987, pp 173–176.
Werman R, Davidoff RA, Aprison MH: Inhibitory action of glycine on spinal neurons in the cat. J Neurophysiol 1968;31:81–95.

A Cloud of Peptides

Dale [1957] described in 1957 how Oliver and Schaefer [1895a, b] discovered the blood pressure-rising effect of suprarenal extracts in 1895, and out of curiosity investigated the posterior pituitary lobe which also contained a blood pressure-rising agent. Howell [1898] described that this agent is mainly concentrated in the nerve fibres of the posterior lobe of the pituitary. The antidiuretic effect of pituitary extracts was discovered in 1913 by two clinicians, Farini in Venice and van der Velde in Düsseldorf [see Tausk, 1970] and by Konschegg and Schuster [1915]. Kamm et al. [1928] showed that the posterior pituitary contains two agents, vasopressin (the antidiuretic hormone) and oxytocin. The structure and synthesis of oxytocin was established by Tuppy in Vienna, and that of both hormones at the same time by du Vigneaud et al. [1953].

Sir Henry Dale had described already in 1906 that a pituitary extract evoked a contraction of the uterus of the cat. As Chester Jones [1976] mentioned, Dale proved once again Pasteur's celebrated phrase: 'Chance only favours the prepared mind.'

Dale [1954] recalled later:

'It is rather remarkable that I seem to have forgotten, in this connection, the active principle so readily extractable from the posterior lobe of the pituitary – the one chapter of endocrinology to which I had made a positive contribution. And my failure is particularly curious, in view of the fact that such extracts had already been in use, for a good many years before insulin, for the successful control of the much rarer and less serious form of diabetes, due to the deficiency of this pituitary lobe, and known as diabetes insipidus.'

This started the discovery of peptidergic neurotransmitters and neurohormones. Neuropeptides represent one large group of neurotransmitters, beside monoamines and related substances, and the amino acids. They are represented directly in the genome and are not produced from amino acids by specific enzymes as are the monoamine transmitters. In the case of monoamine transmitters only the synthesizing enzymes are stored in the genetic code.

The precursor synthesis of neuropeptides is controlled by mRNA. In situ hybridization of their mRNA in tissue slides makes it possible to visualize their presence with the microscope. From the protein coding regions of the gene (exons) mRNA is transcribed and translated to preprohormones. Cleavage of the N-terminal signal peptide produces the prohormone, from which endopeptidases excise peptide fragments. These may be active as such or undergo further posttranslational processing (e.g. amidation, as in tachykinins). Structurally related peptide families, like the 'tachykinin family' with similar biological activities, or the calcitonin-CGRP, or the PACAP-NPY family with quite a different biological activity of the individual compounds have been found. Some peptides are colocalized with other peptides in one neuron, like CGRP with substance P or with monoamines, like VIP with acetylcholine, or NPY with adrenaline.

Apart from monoamines and amino acids, peptides are not reloaded into the vesicles from which they are released, but are cleaved by a great number of peptidases to amino acids.

The peptide receptors are G protein-gated or tyrosine kinase-linked receptors, their actions develop slowly, and therefore in many instances peptides have to be regarded as neuromodulators rather than as neurotransmitters.

The release of neuropeptides may be either continuous or induced briefly by calcium ions.

Peptide receptor agonists in nature are rare, like morphine acting on μ-receptors for endogenous enkephalins. It proved difficult to find peptide analogues with specific and potent agonist or antagonist actions, but in a few cases it was possible like the bradykinin receptor antagonist icatibant, containing unusual amino acids. Rational guidelines to find synthetic nonpeptide receptor antagonists of peptides, comparable to amine antagonists, do not exist. The search for peptide antagonists seems to be comparable to gold washing by testing thousands of substances using a robot.

There were dramatic changes between the periods of time of amines and peptides. No new monoamine transmitters have appeared during the last decades, whereas the number of peptides has been continuously increasing during that time. New extraction and isolation methods were found: the solid-phase synthesis of peptides (Merrifield, Nobel Prize for Chemistry in 1984), the radioimmunoassay to measure them in tissue extracts (Guillemin, Schally, Yalow, Nobel Prize for Physiology or Medicine in 1977), and the immunohistochemistry to enable their localization. Furthermore, highly sensitive and specific assays to measure mRNA have been employed in the elucidation of the regulation of precursor synthesis.

Only peptides which are produced in the cell body of neurons, transported to the nerve terminals, released when these neurons are stimulated, and acting on synaptically adjacent neurons or on adjacent target cells are considered to be neuropeptides. Peptides with endocrine functions, transported with the bloodstream to their targets, are considered peptide hormones and are not included in this book.

References

Chester Jones I: Evolution aspects of the adrenal cortex and its homologues. The Sir Henry Dale Lecture for 1976. J Endocrinol 1976;71:3.

Dale HH: An autumn gleaning; in Occasional Lectures and Addresses. London, Pergamon Press, 1954.

Dale HH: Evidence concerning the endocrine function of the neurohypophysis and its nervous control; in Heller H (ed): The Neurohypophysis. London, Butterworths, 1957, pp 1–9.

Du Vigneaud V, Ressler C, Swan JM, Roberts CW, Katsoyannis PG, Gordon S: The synthesis of an octapeptide amide with the hormone activity of oxytocin. J Am Chem Soc 1953;75:4879–4880.

Howell WMU: The physiological effects of the hypophysis cerebri and infundibular body. J Exp Med 1898;3: 245–258.

Kamm O, Aldrich TB, Grote IW, Rowe LW, Bugbee EB: The active principles of the posterior lobe of the pituitary gland. J Am Chem Soc 1928;50:573–601.

Konschegg A, Schuster E: Über die Beeinflussung der Diurese durch Hypophysenextrakte. Dtsch Med Wochenschr 1915;41:1091.

Oliver G, Schäfer EA: On the physiological actions of the pituitary body and other glandular organs. J Physiol 1895a;18:277–279.

Oliver G, Schäfer EA: The physiological effects of extracts of the suprarenal capsules. J Physiol 1895b;18:230–276.

Tausk M: Pharmakologie der Hormone. Stuttgart, Thieme, 1970.

Substance P

Substance P was discovered by von Euler and Gaddum [1931] and von Euler [1977] described how this happened:

'The discovery of substance P (SP) was unexpected but not wholly fortuitous. As a postgraduate student in 1930 I was allowed to work in H.H. Dale's laboratory in Hampstead, London. At that time acetylcholine (ACh) was of primary interest in the laboratory, and I was given the task of trying to demonstrate a release of ACh from the intestine on vagus stimulation. This led to some experiments in which an intestinal extract was tested for biological activity on the isolated jejunum of the rabbit. Since the contraction observed was not inhibited by atropine and since histamine – another of the favorite substances of the laboratory – did not contract the rabbit's jejunum, it appeared that the effect observed was due to a new or at least unknown principle. Dale suggested that the work should be continued jointly with J.H. Gaddum, senior assistant in the laboratory. As a result the effect observed was systematically investigated, and after some time we were satisfied that the active agent was not identical with any of the compounds then known that had a stimulating action of the gut. A study of extracts from various organs also showed that this substance was present in appreciable quantities in the brain. We also observed that all extracts that contracted the gut also lowered the blood pressure, especially in the rabbit. In order to make quantitative estimations we used a purified standard preparation simply referred to as "P" on the tracings and in the protocols. Some 30 years later, Gaddum wrote, "We concentrated the active substance in the form of a stable dry powder known as preparation P (which probably contained about 15 units/mg). It is impossible to justify the widespread custom, started by Gaddum and Schild, of calling the active principle itself substance P, but it is probably too late to change that now." The active principle is now generally called substance P or SP, somewhat in analogy to PG for prostaglandins, which has the advantage of allowing convenient addition of qualificatory suffixes.'

Gaddum and Schild [1934] excluded adenosine derivatives in this extract, found its acid stability, its solubility in acetone and ethyl alcohol, and described it as a basic compound. Von Euler [1936] removed large proteins from the tissue extracts by boiling at pH 4, followed by salting out the peptides by ammonium sulphate. This stable dry powder, which probably contained 15 units/mg, was used in all the earlier experiments. The peptide nature of the active principle was indicated by inactivation with trypsin.

It is worth remembering what Dale [1933] said:

'The discovery, in artificial extract from an organ or tissue, of a substance which on artificial injection produces a pharmacodynamic effect provides only a first item of presumptive evidence in support of a theory that the action of this substance plays a part in normal physiology. Much more evidence is required before we can attribute clearly defined functions to such a substance, as we can now do in the cases of histamine and acetylcholine. But even where this is already possible, we have still no evidence to justify the assumption that the substance comes naturally into action in the body in the free condition in which we isolate and identify it in the laboratory after various unnatural chemical procedures.'

Not much attention was paid to substance P for the next 20 years. At that time a number of peptides such as the hormones insulin, vasopressin and oxytocin, and other agents like bradykinin or angiotensin were known. But no methods were yet available for their isolation and establishing their structure.

My interest in substance P emerged from other findings (F.L.). Hellauer, the youngest coworker of Otto Loewi in Graz, had published with Loewi two papers in 1940, in which it was shown that the dorsal roots of the spinal cord, in contrast to the ventral roots, do not contain acetylcholine. Long ago, Stricker [1876] had made an unusual observation: when he stimulated the peripheral endings of the cut dorsal roots in a dog, he observed a peripheral vasodilatation, measured by a mercury thermometer between the toes. This was a contradiction to the Law of Bell and Magendie which says that the dorsal roots contain only afferent nerve fibres. Stricker's observation was fully confirmed by Bayliss [1901] and was thereafter mentioned as an unexplained curiosity by the name 'antidromic vasodilatation' in textbooks of physiology.

Dale commented in his Walter Dixon Memorial Lecture (1935):

'When we are dealing with two different endings of the same sensory neuron, the one peripheral and concerned with vasodilatation and the other at a central synapse, can we suppose that the discovery and identification of a chemical transmitter of axon-reflex vasodilatation would furnish a hint as to the nature of the transmission process at a central synapse? The possibility has at least some value as a stimulus to further experiment.'

Having been trained in bioassay in Gaddum's laboratory, I approached the issue of finding the transmitter substance of sensory fibres. I used the isolated guinea pig ileum as a bioassay, prepared the substance P-containing ammonium precipitate and found that it caused contractions in the presence of atropine and an antihistamine. Therefore, I interpreted it as the effect of substance P. I went to the slaughterhouse in order to collect ventral and dorsal roots and made simple extracts. And really, there was at least 10 times more activity in the dorsal as compared to the ventral roots [Lembeck, 1953a]. The dorsal roots had the activity of 4 units of substance P/mg wet tissue, thus equalling roughly 20 ng substance P. I had not been aware before that the guinea pig ileum, used for the first time as a substance P bioassay, reacted to such a small amount. A proteolytic destruction of the activity of the dorsal and ventral roots could also be shown, which was a strong argument for substance P. As a second bioassay I used the fall in blood pressure of the atropinized rabbit, but there was no difference in the activity.

In parallel to these investigations, Kopera and Lazarini [1953] studied the distribution of substance P in the bovine and feline brain. Simple extracts were prepared and tested on the guinea pig ileum under atropine and an antihistamine. A high amount of substance P in the substantia nigra was noted. These results were confirmed and extended by investigations of Pernow [1953], Zetler and Schlosser [1955], and others [(for a review, see Pernow, 1983].

All these findings evoked a certain amount of interest in substance P and led to a first Symposium on Substance P, held on June 9 and 10, 1961, in Sarajevo. At the end of this symposium, all participants agreed that further progress would depend on the expected isolation of substance P. Pernow [1953] had already found that substance P could be absorbed on alumina and eluted with falling concentrations of aqueous methanol; in this way the activity was increased from a few biological units per milligram to about 3,000. Franz et al. [1961] of Basel reported at the symposium that they had obtained an activity of 30,000–35,000 units/

mg by chromatography and electrophoresis. They all had almost reached the goal, but the essential result came from somewhere else.

Erspamer had found that the salivary gland of a mollusc, *Eledone moschata*, contains besides serotonin the peptide eledoisin, and the skin of a South American frog, *Physalemus fuscomaculata,* the peptide physalaemin. Both peptides occur in large amounts and, therefore, their isolation and the elaboration of their chemical structure was no problem. I once asked him about the motivation behind these experiments. His reply was short and precise: 'It was just serendipity, nothing else.' The strategy of his work was to test the activity of eledoisin, physalaemin and substance P on many bioassay preparations in vitro and in vivo. Ten of the amino acids in physalaemin were present in the not yet completed analysis of substance P by the researchers in Basel. In a paper with Bertaccini et al. [1965] we find the following:

'But the door still remained closed for several years. I think the next important step emerged from the Mediterranean Sea, like Aphrodite, but in the shape of a mollusk, *Eledone moschata*. Erspamer discovered eledoisin in its salivary glands and among so many other interesting peptides, shortly thereafter, physalaemin in the skin of a South American frog. The isolation and careful pharmacological analysis of these peptides led to a statement by Erspamer in 1965 [Bertaccini et al., 1965] which is worth quoting:

"It may be seen that as many as ten of the amino acids found in substance P are also present in eledoisin, and eight in physalaemin. The only amino acid lacking in the molecule of substance P is apparently the tremendously important methioninamide. The suspicion seems to be justified that the labile methioninamide residue has escaped the attention of research workers, who have isolated and studied substance P. We would suggest that this possibility be checked".'

At this time, I compared eledoisin and physalaemin with substance P in their effects on the salivary secretion. Together with Starke [Lembeck and Starke, 1968] we measured the salivary secretion after intravenous injection of the substance P preparation in chickens, rats and dogs, the species used by Bertaccini et al. [1965] and observed a pronounced increase in salivary secretion by substance P .

On May 4, 1967, I made an important discovery in an unusual place. Sitting in a comfortable intercity train, reading the *Frankfurter Allgemeine Zeitung*, I found a short note about the isolation of a 'salivary secretion stimulating factor' called 'sialogen' in peptide extracts from the hypothalamus by Leeman and Hammerschlag [1967], presented at a 'Federation Meeting'. I had the feeling that I was the only really fascinated reader of this message, written by an unknown reporter. After my return to Tübingen, I sent Leeman all the results on salivation by air mail. Samples were exchanged and there was soon enough evidence that sialogen and substance P were identical.

Susan Leeman worked in the field of endocrinology, specifically in the neurohumoral control of ACTH. The neurohormonal control of the ACTH secretion is based on the release of the corticotropin-releasing factor (CRF), and Leeman's project was to purify this factor [Leeman et al., 1962]. The presence of CRF in a crude hypothalamic extract was verified and the extract was subjected to gel exclusion chromatography on Sephadex G-75 as the next step in a purification procedure. Each fraction from the column had to be injected into an anaesthetized rat to measure the biological activity. Material from certain fractions rapidly caused a copious secretion of saliva. Michael Chang joined the project and they moved from CRF to the isolation of the newly discovered hypothalamic peptide. The initial large-scale tissue extraction (the extract from 200 bovine hypothalami was applied to one column!) outgrew the facilities of the laboratory. Further steps were ion exchange chromatography and paper electrophoresis [Chang and Leeman, 1970]. The amino acid sequence of substance P was published in 1971 [Chang et al., 1971]. Tregear et al. [1971] were able to synthesize the peptide and to produce

Tyr^8-substance P which enabled the labelling with ^{125}I and thus permitted the development of a radioimmunoassay [Powell et al., 1973]. Studer et al. [1973] showed that substance P in the gut was identical with that in brain.

This story is strongly reminiscent of what Gaddum wrote in 1954, and what will, hopefully, remain true in the future:

'The most interesting discoveries (in pharmacology) are the unexpected ones, and in spite of the fact that research is much more organized today than it was once, unexpected discoveries are still made.'

The availability of synthetic substance P and of the substance P radioimmunoassay dramatically changed the scenery, and a wave of investigations covered all fields of biomedical research [see Pernow, 1983]. New methods allowed a much better insight into the distribution of substance P in tissues and conclusions about its function. At the Nobel Symposium 37 under the title 'Substance P', organized by U.S. von Euler and Bengt Pernow in Stockholm in 1975, the dominant role of the immunological methods was visible. Thomas Hökfelt presented the distribution of substance P in the central (CNS) and peripheral nervous system.

Masanori Otsuka and the Action of Substance P in the Spinal Cord

At the International Congress of Pharmacology in San Francisco in 1972 I met Marthe Vogt, who asked whether I knew Masanori Otsuka. I denied. 'But he has confirmed your results on substance P from 1953', she said and brought me and Otsuka together.

Otsuka found that the dorsal roots contain 20 times more substance P than the ventral roots, and he first observed with the use of a bioassay that the content of substance P in the dorsal horn declined after section of the dorsal roots; this means that substance P is produced in the dorsal root ganglion and transported by the central fibres of afferent neurons into the dorsal horn [Otsuka et al., 1972]. I returned from California with a feeling like that of the gold-digging people. Pernow also confirmed it: the rat ventral horn contains 134 ± 33 pmol/g, the dorsal horn 1,070 ± 160 pmol/g of wet tissue substance P [see Pernow, 1983].

Konishi and Otsuka [1974a] investigated the effects of substance P and other peptides on the isolated spinal cord of the frog. Thereafter they used the spinal cord of newborn rats [Konishi and Otsuka, 1974b]. The dorsal root was placed in a suction electrode which was stimulated; the ventral root was placed in a recording suction electrode. The depolarization in the ventral root was measured after the addition of substance P and *L*-glutamate, and the activity of substance P was estimated to be about 200 times higher than that of glutamate, which was regarded as the leading candidate for the excitatory transmitter in the spinal cord [Curtis and Johnston, 1974]. Not earlier than in 1994 [Ueda et al., 1994] was the co-release of substance P and glutamate shown. Yanagisawa et al. [1984] developed the isolated spinal cord tail preparation of the newborn rat, by which they could show capsaicin-induced nociceptive responses [for the neurotransmitter functions of mammalian tachykinins, see also the review by Otsuka and Yoshioka, 1993].

Further Tachykinins and Many More Peptidergic Neurons

The word 'tachykinin' was a creation of Erspamer. Eledoisin, physalaemin and substance P caused faster contractions than bradykinin on isolated smooth muscle preparations; to define this difference at a time when none of these peptides was isolated, he created these descriptive names which are still used today. Two peptides closely related to substance P, neurokinin A and B, were discovered in mammalians, and many further tachykinins were isolated in lower animals.

The immunohistochemical technique, including the immunofluorescence technique, the peroxidase technique, the very sensitive peroxidase-antiperoxidase technique and the 'double staining techniques' allowed the identification of more than 20 peptides in neurons of the brain, spinal cord and in the periphery. In several cases peptides occur together with a 'classical' transmitter in the same neuron [Hökfelt et al., 1980]. Calcitonin gene-related peptide (CGRP) was an unexpected finding and remarkable, because its distribution and functions are closely related to those of substance P [for a review, see Wimalawansa, 1996]. Somatostatin was also found concomitantly with substance P [Gamse et al., 1981]. The concentration of these and many other peptides in the CNS is about 1,000 times lower than that of monoamines and 100,000 times less than that of amino acids. But peptides may activate their receptors at a much lower concentration than classical transmitters. Classical transmitters are not only synthesized in the cell body, but also at the nerve endings and a reuptake mechanism exists. Peptides are produced only in the cell body, carried by the axonal transport to the nerve terminals where no reuptake mechanism seems to exist. Classical transmitters usually cause a rapid response of short duration, whereas the peptides are responsible for a long-lasting effect. It has to be emphasized that the new findings of Hökfelt and many others provided a completely new view of the afferent, autonomic and CNS. The findings also include new aspects in the organization of neuronal pathways within the CNS. Finally the coexistence of peptides with classical transmitters expanded the original principle of Dale (i.e. one neuron-one transmitter) to a system which seems to be adequate for the great variety of neuronal functions, especially within the CNS [for a review, see Hökfelt et al., 1980].

Nicolas Jancsó and Capsaicin

In 1967 I met Nicolas Jancsó in Szeged. I knew Jancsó's work on the extinction of pain by capsaicin [Jancsó and Jancsó-Gábor, 1959]. I invited him to Tübingen for a seminar. Jancsó explained in the seminar his simple, but conclusive experiments, and I was probably the only one in the audience who was convinced that he was right. The work of Jancsó et al. [1967] concluded the following:

(1) By antidromic electrical stimulation of the sensory nerves (saphenous or trigeminal) of rats, the following signs of an inflammatory response could be elicited: arteriolar vasodilatation, enhancement of vascular permeability, protein exudation, fixation of injected colloidal silver onto the walls of venules and, later, their storage in histiocytes.

(2) The inflammatory response induced by electrical stimulation could not be altered by parenterally administered atropine, physostigmine, hexamethonium, phentolamine, dibenamine, propranolol, promethazine, chloropyramine, or methysergide.

(3) After the degeneration of the sensory nerve, capsaicin, xylene, mustard oil and ω-chloroacetophenone did not evoke inflammation. Hence, these substances induce inflammation purely or dominantly through the involvement of sensory nerves.

(4) Capsaicin desensitization inhibited the signs of inflammation induced both by antidromic stimulation of the sensory nerve and by orthodromic stimulation of pain-sensitive nerve terminals with irritants.

(5) The experiments suggest that a mediator substance, a neurohumor, is released by orthodromic or antidromic stimulation of pain-sensitive nerve terminals and that this substance is responsible for the signs of inflammation produced by some substances.

The question was whether the proposed 'neurohumor' was indeed substance P. The paper by Gamse et al. [1980] supplied a clear answer: we checked whether antidromic vasodilatation, so far only shown by Bayliss [1901] in the dog, could be reproduced in the rat hind leg. By means of a handmade drop recorder, we found that the venous outflow from the hind leg was increased by a 1-min stimulation of the saphenous nerve for about 10 min, comparable to the result of Bayliss [1901] in the dog. In capsaicin-pretreated rats an almost complete inhibition was observed [Lembeck and Holzer, 1979]. Jancsó had demonstrated that the 'neurogenic inflammation' consisted of vasodilatation and plasma extravasation. Substance P also evoked a plasma extravasation by histamine release from nearby mast cells, whereas CGRP did not. These events seemed to be closely related to the 'nocifensor system', which Sir Thomas Lewis had found by studies on the human skin [see Lembeck, 1985]. It consisted of a peripheral neurogenic vasodilatation and plasma extravasation (flare and wheal response), providing a defence mechanism which removes exogenous or endogenous toxic material by increase of blood flow and lymph drainage [Lundberg and Saria, 1983; see also Lembeck, 1988]. When selective substance P (NK_1) receptor antagonists became available, it was definitively proven that substance P is the mediator of neurogenic plasma extravasation [Lembeck et al., 1992].

Peptide and Non-Peptide Transmitters in Primary Afferent Neurons

Of the primary afferent neurons, Aβ fibres detect innocuous stimuli applied to skin, muscle and joints. Stimulation of Aβ fibres can reduce pain, as it occurs when you activate them by rubbing your hand. Thinly myelinated Aδ fibres and the most slowly conducting C fibres transmit the rapid, acute, sharp pain, and the delayed, more diffuse, dull pain, respectively. Aδ nociceptors respond to intense mechanical stimuli and to intense heat. Most C fibres are polymodal, responding to noxious thermal, chemical and mechanical stimuli. Others are mechanically insensitive, but respond to noxious heat. Most C fibre nociceptors respond to chemical stimuli like capsaicin or acid. Tissue injury might sensitize so-called 'silent' receptors. The peptidergic group of unmyelinated C fibres contains tachykinins (substance P and neurokinin A) and expresses trkA (high-affinity tyrosine kinase) receptor for nerve growth factor. The second population expresses P2X receptors, specific ATP-gated ion channels. Further peptides found in C fibres include CGRP, concomitantly released with substance P [for a review, see Holzer, 1992], somatostatin, neurotensin, vasoactive intestinal polypeptide and galanin, with much less-defined functions. Besides peptides, glutamate is most likely the predominant excitatory neurotransmitter in all primary afferent neurons.

In contrast to vision, olfaction or taste, sensory nerve endings that detect painful stimuli ('nociceptors') are not localized to a particular anatomical structure, but are dispersed over the body, innervating skin, muscle, joints and internal organs. The transmission of pain has attracted the greatest interest in recent years. The vanilloid receptor VR1 was cloned and functionally characterized; it is activated by heat, capsaicin and acid [for a review, see Szallasi and Blumberg, 1999]. The VRL-1 receptor, expressed in a subset of medium- to large-diameter myelinated neurons, is activated by noxious heat. VR1 and VRL-1 belong to a larger family of transient receptor potential channels [for a review, see Julius and Basbaum, 2001].

Primary afferent C fibres are stimulated at their peripheral endings by many exogenous and endogenous substances, like capsaicin, mustard oil, xylol and other compounds which easily penetrate the skin. Following stimulation, substance P is released from the peripheral terminals of sensory neurons, as shown by the vasodilatation and plasma extravasation due to antidromic electrical nerve stimulation. Plasma extravasation may be augmented by the release of histamine from mast cells, specifically by basically charged peptides, like substance P, but not by CGRP. This recalls the 'nocifensor system' claimed by Sir Thomas Lewis. He showed the cutaneous 'wheel and flare response' to be of neurogenic origin, absent after chronic denervation, in clinical studies on the human skin. As the antidromic vasodilation can only be evoked by nerve stimulation, it cannot be regarded as physiological. But under physiological conditions two actions occur, firstly the transmitter release at the peripheral terminals, and secondly the neuronal conduction to the central terminals of the dorsal root fibres within the CNS. At this site, the release of substance P takes place, and the amount of substance P is increased under inflammatory conditions in the periphery. This is, therefore, a nociceptive pathway. But the capsaicin-sensitive C fibre afferents are not only involved in the signalling of pain. They represent much more the afferent part of the autonomic nervous system (which we erroneously regard as being only efferent!).

Alarming messages recruiting one or the other defence reaction usually arise from higher brain centres, but the initiating information is signalled to the brain from the periphery. The signal might activate a single, unconscious reaction or induce more than one of the possible defence reactions. The peripheral pathways which signal the alarming event to the CNS seem mainly to be afferent neurons which share certain common properties: (1) they are small-diameter, unmyelinated C fibres, (2) they synthesize substance P and other peptides and transport them to the terminals where they are released by nervous impulses, and (3) these neurons are capsaicin-sensitive, i.e. a small dose of capsaicin causes the release of the peptide, and systemic treatment with a very large dose of capsaicin causes these neurons to lose their function or to degenerate. The loss of function after systemic capsaicin treatment can, therefore, be regarded as evidence for the involvement of these afferent fibres in the response. It does not allow to neurochemically define the type of neuron that is involved. Therefore, it has been agreed to speak of 'capsaicin-sensitive neurons'. This limitation is essential because other C fibres exist which are not capsaicin-sensitive: e.g. cold fibres [Holzer-Petsche and Lembeck, 1984] and osmosensitive fibres in the hepatic portal vein [Stoppini et al., 1984].

Involvement of Substance P in Autonomic and Neuroendocrine Reflexes

Orthodromic conduction and transmitter release from capsaicin-sensitive neurons can evoke several reflex responses.

(1) At the spinal level it induces the scratching syndrome in mice [Piercey et al., 1985].

(2) Cardiovascular reflexes mediated by capsaicin-sensitive afferent fibre stimulation are set in action either by a stimulation or a withdrawal of the noradrenergic vasoconstrictor tone [Juan and Lembeck, 1974; Donnerer and Lembeck, 1988].

(3) Stimulation by heat induces the heat loss reaction in rats [Donnerer and Lembeck, 1983; Cormareche-Leyden et al., 1985].

(4) The micturition reflex in rats caused by distension of the urinary bladder involves capsaicin-sensitive afferents [Holzer-Petsche and Lembeck, 1984; Maggi and Meli, 1988] whose activation might depend on ATP.

Endocrine regulations influenced by capsaicin-sensitive afferents and possibly comediated by tachykinins have also been revealed.

(1) The hepatic portal vein of rats contains osmoreceptors and glucoreceptors [Stoppini et al., 1984] which are sensitive to hypoglycaemia and signal this information to hypothalamic centres via capsaicin-sensitive afferents, which in turn leads to release of adrenaline from the adrenal medulla to upregulate the blood glucose concentration [Amann and Lembeck, 1986; Donnerer, 1988].

(2) The release of ACTH from the pituitary can be evoked either by emotional or various peripheral 'stressors'. Capsaicin-sensitive neurons seem to convey such peripheral signals to the CNS. Cold exposure of rats is known to release ACTH; in capsaicin-treated rats, this response is abolished. Other painful peripheral stimuli like intraperitoneal application of formalin induce the release of ACTH; the effect of these treatments is abolished in capsaicin-treated rats. Emotional stress, induced by restraint conditions, which is not conveyed by capsaicin-sensitive afferents but by other central path-

ways, remains unchanged in capsaicin-treated rats [Amann and Lembeck, 1987; Donnerer and Lembeck, 1988]. Capsaicin-treated rats have normal levels of plasma ACTH, which means a normal function of the endocrine regulation of glucocorticoids. Only the stimulation of the immediate release of ACTH, emerging from afferent neuronal signals, is under the influence of capsaicin-sensitive afferents.

(3) The fertilized egg cell develops to a blastocyte during the transport into the rat uterine cavity within 3 days. In the meantime endocrine influences have transformed the uterine mucosa to the decidua, an essential requirement for the nidation. These endocrine adaptations via hypothalamus and pituitary are set into action by afferent messages emerging from the stimulation of vagina and cervix during copulation in the rat. These messages run via capsaicin-sensitive neurons to the CNS [Traurig et al., 1984].

These few examples of autonomic reflexes and endocrine regulations present evidence that the afferent part is mediated by capsaicin-sensitive neurons. It is known that the nervus vagus or the nervus splanchnicus, described in textbooks as 'efferent', contain a high amount of afferent fibres.

Tachykinin Genes, Precursors and Receptors

A new era in tachykinin research began in the 1980s with the application of recombinant DNA technology. Nakanishi and his coworkers (1983–1986) opened the way to investigate the biosynthesis of substance P and first isolated its gene and precursor protein containing 384 amino acid residues [Nawa et al., 1983; Kawaguchi et al., 1986].

The *Xenopus* oocyte expression system combined with electrophysiological measurements was the key to the molecular structure and function of the three tachykinin receptors. Intranuclear injection of oocytes with cDNA encoding the receptor or injection of cRNA into the cytoplasm of an oocyte produces a functional, foreign receptor-channel complex in the membrane of the oocyte. After application of its specific ligand, an electrophysiological response can be recorded. It was shown that the preferred receptors for substance P, neurokinin A and neurokinin B are encoded by different mRNAs. The investigation of the relative activities on various test preparations showed that the functional site of the NK_1 receptor is predominantly activated by substance P, that of the NK_2 receptor by neurokinin A, and that of the NK_3 receptor by neurokinin B [see Henry et al., 1987]. The membrane topology of all three receptor molecules is that of rhodopsin-type receptors, consisting of seven hydrophobic membrane-spanning domains with an extracellular amino terminus and a cytoplasmic carboxyl terminus [Nakanishi, 1991].

Co-Localization of Substance P with Classical Transmitters

Until the 1970s it had been a widespread belief that a neuron contains and releases only one transmitter according to Dale's concept of one neuron-one transmitter. Hökfelt et al. [1978] showed the co-existence of substance P and serotonin in the same neuron, a finding later confirmed by Cuello et al. [1982]. There are numerous examples of the co-existence of substance P and serotonin not only within the CNS, but also in the periphery. Neurokinin A and serotonin co-exist in carcinoid tumours [Bergström et al., 1995]. When Lembeck [1953b] first found serotonin in a carcinoid tumour, he noted another gut-stimulating compound with a different Rf value. By desensitization with large doses of substance P and serotonin, respectively, he found an indication for the presence of both compounds. But the methods available at that time were not sufficient for a clear proof and he mentioned this finding only in the discussion but not in the summary of the paper.

Numerous examples of the co-existence of peptides with monoamines and amino acid neurotransmitters have now been established. The functional significance of more than one transmitter is still not fully understood at present. If a neuron stores and releases a fast-acting transmitter (e.g., an amino acid) and a slowly acting transmitter (e.g., a peptide), then upon excitation it will produce a fast excitatory postsynaptic potential (EPSP) or inhibitory postsynaptic potential (IPSP) and in addition slow EPSPs and IPSPs in postsynaptic cells [Otsuka and Yanagisawa, 1987]. For example, if GABA and substance P are co-released, a short-lasting inhibition and a prolonged excitation will be produced [Otsuka and Yoshioka, 1993]. In some circumstances, multiple transmitters released from a nerve terminal may diffuse across some distance and selectively act on cells equipped with the respective receptors, called 'chemically addressed transfer of information' [Iversen, 1986].

An interesting issue is the co-release of substance P and glutamate. De Biasi and Rustioni [1988] first showed the co-existence of glutamate and substance P in dorsal root ganglion neurons and at their central terminals. NMDA receptor blockade of afferent nerve-induced autonomic reflexes was shown by Donnerer and Amann [1994]. Ueda et al. [1994] found that glutamate is released from capsaicin-sensitive primary afferent fibres in the rat with online continuous monitoring of glutamate. The inhibition by the NMDA receptor antagonist MK-801 of the central transmission of reflexes induced by capsaicin-sensitive afferents indicates the essential role of glutamate in these reflex responses, whereas substance P antagonists had no effect [Juránek and Lembeck, 1996]. A facilitation of the glutamate release by co-released substance P was assumed [Juránek and Lembeck, 1997].

The life cycle of substance P ends by internalization of the receptor-bound peptide and enzymatic breakdown. The degradation of substance P is difficult to define, as several organs (kidney, spleen, liver and intestine homogenates) have a high capacity to inactivate it. One has to differentiate between elimination of substance P released into the circulation and substance P destroyed locally such as in the brain. By using the rat salivary response as a bioassay, Lembeck et al. [1978] showed that the highest degree of elimination of substance P occurs in the liver and hind limbs, followed by the kidney. More than 80% of substance P infused into the portal vein is degraded by the liver.

A number of cytosolic and membrane-bound peptidases capable of degrading substance P has been extracted from the brain. A partly purified neutral endopeptidase has first been described by Benuck and Marks [1975]. Degradation of substance P by a neutral metallopeptidase system of synaptosomal fractions of brain also exists [Berger et al., 1979]. Neutral endopeptidase subsequently turned out to significantly inhibit the transmitter action of substance P.

Substance P and Neurokinin A in the Gut

Although substance P was discovered in gut and brain in 1931, by a fall in blood pressure and by the contraction of the rabbit ileum, nobody asked what its physiological role might be in the latter. Acetylcholine and noradrenaline were accepted as gastrointestinal neurotransmitters, but about a dozen other compounds were known to contract intestinal smooth muscle, such as histamine, oxytocin, later on serotonin, bradykinin and other compounds. The first hint that substance P is involved in neuronal functions was the finding of Ehrenpreis and Pernow [1953] that the aganglionic, inactive part of the rectosigmoid in Hirschsprung's disease contained significantly less substance P than control pieces, whereas the proximal, hyperactive part showed normal levels of substance P. Substance P was found in all parts of the intestine of animal species and man. The guinea pig ileum was later used as a bioassay for substance P, since it is particularly sensitive to substance P. The peristaltic reflex method of Holzer and Lembeck [1979] on the guinea pig ileum offered experimental possibilities to investigate the action of substance P on propulsive motility.

The essential breakthrough came from immunohistochemistry. Two fundamental types of neurons supply the digestive tract: an extensive population of neurons contained within the gut wall (intrinsic neurons) and an extrinsic innervation by cholinergic and noradrenergic autonomic fibres as well as by substance P-containing sensory neurons from dorsal root ganglia and the vagus nerve.

The substance P nerve cells in the myenteric ganglia issue a very dense network of varicose nerve fibres to local myenteric ganglia, as well as fibres to the longitudinal and circular muscle layers, to the submucous ganglia and to the mucosa. The number of the myenteric plexus substance P cell bodies has been underestimated. Only 2.8–3.5% of the myenteric neurons contained substance P, but in segments of the intestine in a culture medium with colchicine for 24 h, about 20% of the total population are substance P-containing neurons. The proportion of substance P neurons in the submucous plexus is about 11%. In the human intestine, the substance P neurons have a similar distribution as compared to the guinea pig [Holzer and Holzer-Petsche, 1997a]. The extrinsic sensory substance P neurons are fewer than the intrinsic neurons. The extrinsic neurons are capsaicin-sensitive whereas the intrinsic substance P neurons are not. They sense intestinal pain (colic) or they induce reflexes (vomiting) or diarrhoea to expel toxic material [Holzer and Holzer-Petsche, 1997b].

The enteric nervous system, already separated from the autonomic nervous system by Langley [1932], consists of a large number of neurons, in the range of the number of neurons within the spinal cord. Immunohistochemistry led to the detection of quite a number of active compounds, besides those mentioned above, such as neuropeptide Y, somatostatin, cholecystokinin, VIP, and others [for a review, see Holzer and Holzer-Petsche, 1997a, b].

Tachykinin Receptor Antagonists: The Role of Substance P Revisited

Antagonists of substance P, acting specifically like atropine on muscarinic acetylcholine receptors, were an old dream of pharmacologists. The attempts to find such a compound on a simple preparation like the guinea pig ileum had not been successful. A new way was chosen by Snider et al. [1991] and led to the discovery of the first specific non-peptide antagonist CP-96,345. It was found by the screening of thousands of compounds on the interference with substance P binding. The antagonist was specifically active on the fall in blood pressure induced by injection of substance P and neurokinin A, but not on that by VIP or CGRP. It inhibited the plasma protein extravasation induced by mustard oil in rats, when injected intravenously or given orally.

We probably made an error. The substance P antagonist completely inhibited the effects of substance P released by antidromic nerve stimulation or capsaicin from the peripheral terminals of primary afferent fibres. We knew that substance P is also released from the central terminals of these fibres in the substantia gelatinosa of the dorsal horn of the spinal cord, as shown in vitro on spinal cord slices of the rat [Gamse et al., 1979] and in vivo in the cat [Duggan et al., 1987]. We assumed that several reflexes and endocrine regulations, which were blocked following capsaicin treatment, should also be blocked by the substance P antagonist. But we did not find an action of the substance P antagonist. At the same time the co-release of substance P and glutamate was shown [Ueda et al., 1994]. All the centrally mediated effects evoked or blocked by capsaicin were shown to be inhibited by the NMDA antagonist MK-801, therefore being mediated by glutamate [Juránek and Lembeck, 1997]. The question remains what the physiological effect of substance P at these synapses might be. In the meantime many new and better antagonists blocking NK_1 receptors have been synthesized or isolated from plants, e.g. Compositae [Yamamoto et al., 2002].

In summary, it can be stated that decades of intense research in the field of substance P and its receptors have led to significant advances in the field of physiology, pathophysiology and pharmacology; yet great pharmacological opportunities are still in the future.

References

Amann R, Lembeck F: Capsaicin-sensitive afferent neurons from peripheral glucose receptors mediate the insulin-induced increase in adrenaline secretion. Naunyn Schmiedebergs Arch Pharmacol 1986;334:71–76.

Amann R, Lembeck F: Stress-induced ACTH release in capsaicin treated rats. Br J Pharmacol 1987;90:727–731.

Bayliss WM: On the origin from the spinal cord of the vasodilator fibres of the hind limb, and on the nature of these fibres. J Physiol Lond 1901;26:173–209.

Benuck M, Marks N: Enzymatic inactivation of substance P by a partially purified enzyme from rat brain. Biochem Biophys Res Commun 1975;65:153–160.

Berger H, Fencher K, Albrecht E, Niedrich H: Substance P: in vitro inactivation by rat brain fractions and human plasma. Biochem Pharmacol 1979;28:3173–3180.

Bergström M, Theodorsson E, Norheim I, Öberg K: Immunoreactive tachykinins in 24-h collections of urine from patients with carcinoid tumors: characterization and correlation with plasma concentrations. Scand J Clin Lab Invest 1995;55:679–689.

Bertaccini G, Cei JM, Erspamer V: The action of physalaemin on the systemic arterial blood pressure of some experimental animals. Br J Pharmacol 1965;25:380–391.

Chang MM, Leeman SE: Isolation of a sialogogic peptide from bovine hypothalamus and its characterization as substance P. J Biol Chem 1970;245:4784–4790.

Chang MM, Leeman SE, Niall HD: Amino-acid sequence of substance P. Nat New Biol 1971;232:86–87.

Cormareche-Leyden M, Shimada SG, Stitt JT: Hypothalamic thermosensitivity in capsaicin-desensitized rats. J Physiol 1985;363:227–236.

Cuello AC, Priestley JV, Matthews MR: Localization of substance P in neuronal pathways; in Porter R, O'Connor M (eds): Substance P in the Nervous System. London, Pitman, 1982, pp 55–83.

Curtis DR, Johnston GA: Amino acid transmitters in the mammalian central nervous system. Ergeb Physiol 1974;69: 97–188.

De Biasi S, Rustioni A: Glutamate and substance P coexist in primary afferent terminals in the superficial laminae of the spinal cord. Proc Natl Acad Sci USA 1988;85: 7820–7824.

Donnerer J: Reflex activation of the adrenal medulla during hypoglycemia and circulatory dysregulations is regulated by capsaicin-sensitive afferents. Naunyn Schmiedebergs Arch Pharmacol 1988;338:282–286.

Donnerer J, Amann R: Effect of NMDA receptor blockade on afferent nerve stimulation-induced autonomic reflexes. Naunyn Schmiedebergs Arch Pharmacol 1994; 349:R85.

Donnerer J, Lembeck F: Heat loss reaction to capsaicin through a peripheral site of action. Br J Pharmacol 1983; 79:719–723.

Donnerer J, Lembeck F: Neonatal capsaicin treatment of rats reduces ACTH secretion in response to peripheral neuronal stimuli but not to centrally acting stressors. Br J Pharmacol 1988;94:647–652.

Donnerer J, Zhao Y, Lembeck F: Effects of clonidine and yohimbine on a c-fibre-evoked blood pressure reflex in the rat. Br J Pharmacol 1988; 94: 848–852.

Duggan AW, Morton CR, Zhao ZQ, Hendry IA: Noxious heating of the skin releases immunoreactive substance P in the substantia gelatinosa of the cat: a study with antibody microprobes. Brain Res 1987;403:345–349

Ehrenpreis T, Pernow P: On the occurrence of substance P in the rectosigmoid in Hirschsprung's disease. Acta Physiol Scand 1953;27:380–388.

Franz J, Boissonnas RE, Stürmer E: Isolierung von Substanz P aus Pferdedarm und ihre biologische und chemische Abgrenzung gegenüber Bradykinin. Helv Chim Acta 1961;44:881–883.

Gaddum JH, Schild H: Depressor substances in extracts of intestine. J Physiol Lond 1934;83:1–14.

Gamse R, Holzer P, Lembeck F: Decrease of substance P in primary afferent neurons and impairment of neurogenic plasma extravasation by capsaicin. Br J Pharmacol 1980; 68:207–213.

Gamse R, Leeman SE, Holzer P, Lembeck F: Differential effects of capsaicin on the content of somatostatin, substance P, and neurotensin in the nervous system of the rat. Naunyn Schmiedebergs Arch Pharmacol 1981;317: 140–148.

Gamse R, Molnar A, Lembeck F: Substance P release from spinal cord slices by capsaicin. Life Sci 1979;25:629–636.

Henry JL, Couture R, Cuello AC, Pelletier, Quirion R, Regoli D: Substance P and Neurokinins. New York, Springer, 1987.

Hökfelt T, Johansson O, Ljungdahl A, Lundberg JM, Schultzberg M: Peptidergic neurons. Nature 1980;284: 515–521.

Hökfelt T, Ljungdahl A, Steinbusch H, Verhofstad G, Nilsson E, Brodin E, Pernow P, Goldstein M: Immunohistochemical evidence of substance P-like immunoreactivity in some 5-hydroxytryptamine-containing neurons in the rat central nervous system. Neuroscience 1978;3: 517–538.

Holzer P: Peptidergic sensory neurons in the control of vascular functions: mechanisms and significance in the cutaneous and splanchnic vascular beds. Res Physiol Biochem Pharmacol 1992;121:49–146.

Holzer P, Holzer-Petsche U: Tachykinins in the gut. I. Expression, release and motor function. Pharmacol Ther 1997a;73:173–217.
Holzer P, Holzer-Petsche U: Tachykinins in the gut. II. Roles in neural excitation, secretion and inflammation. Pharmacol Ther 1997b;73:219–263.
Holzer P, Lembeck F: Effect of neuropeptides on the efficiency of the peristaltic reflex. Naunyn Schmiedebergs Arch Pharmacol 1979;307:257–264.
Holzer-Petsche U, Lembeck F: Systemic capsaicin treatment impairs the micturition reflex in the rat. Br J Pharmacol 1984;83:935–941.
Iversen LL: Introduction; in Iversen LL, Goodman E (eds): Fast and Slow Chemical Signalling in the Nervous System. Oxford, University Press, 1986, pp xi–xiii.
Jancsó N, Jancsó-Gábor A: Dauerausschaltung der chemischen Schmerzempfindlichkeit durch Capsaicin. Naunyn Schmiedebergs Arch Exp Pathol Pharmakol 1959; 236:142–145.
Jancsó N, Jancsó-Gábor A, Szolcsányi J: Direct evidence for neurogenic inflammation and its prevention by denervation and by pretreatment with capsaicin. Br J Pharmacol Chemother 1967;31:138–151.
Juan H, Lembeck F: Action of peptides and other algesic agents on paravascular pain receptors of the isolated perfused rabbit ear. Naunyn Schmiedebergs Arch Pharmacol 1974;283:151–164.
Julius D, Basbaum AI: Molecular mechanisms of nociception. Nature 2001;413:203–210.
Juránek I, Lembeck F: Evidence for the participation of glutamate in reflexes involving afferent, substance P-containing nerve fibres in the rat. Br J Pharmacol 1996;117: 71–78.
Juránek I, Lembeck F: Afferent C-fibres release substance P and glutamate. Can J Physiol Pharmacol 1997;75: 661–664.
Kawaguchi Y, Hoshimaru M, Nawa H, Nakanishi S: Sequence analysis of cloned cDNA for rat substance P precursor: existence of a third substance P precursor. Biochem Biophys Res Commun 1986;139:1040–1046.
Konishi S, Otsuka M: The effects of substance P and other peptides on spinal neurons of the frog. Brain Res 1974a;65:397–410.
Konishi S, Otsuka M: Excitatory action of hypothalamic substance P on spinal motoneurones of newborn rats. Nature 1974b;252:734–735.

Kopera H, Lazarini W: Zur Frage der zentralen Übertragung afferenter Impulse. IV. Die Verteilung der Substanz P im Zentralnervensystem. Arch Exp Pathol Pharmakol 1953; 219:214–222.
Langley JN: Antidromic action. J Physiol Lond 1932;57: 428–446.
Leeman SE, Glenister DW, Yates FE: Characterization of a calf hypothalamic extract with adrenocorticotropin-releasing properties: evidence for a central nervous system site for corticosteroid inhibition of adrenocorticotropin release. Endocrinology 1962;70:249–262.
Leeman SE, Hammerschlag R: Stimulation of salivary secretion by a factor extracted from hypothalamic tissue. Endocrinology 1967;81:803–810.
Lembeck F: Zur Frage der zentralen Übertragung afferenter Impulse. III. Mitteilung. Das Vorkommen und die Bedeutung der Substanz P in den dorsalen Wurzeln des Rückenmarks. Arch Exp Pathol Pharmakol 1953a,219: 197–213.
Lembeck F: 5-Hydroxytryptamine in a carcinoid tumor. Nature Lond 1953b;172:910.
Lembeck F: Sir Thomas Lewis's nocifensor system, histamine and substance P-containing primary afferent nerves; in Bousfield D (ed): Neurotransmitters in Action. Amsterdam, Elsevier, 1985, pp 173–179.
Lembeck F: The 1988 Ulf von Euler lecture. Substance P: from extract to excitement. Acta Physiol Scand 1988; 133:435–454.
Lembeck F, Donnerer J, Tsuchiya M, Nagahisa A: The nonpeptide tachykinin antagonist, CP-96,345, is a potent inhibitor of neurogenic inflammation. Br J Pharmacol 1992;105:527–530.
Lembeck F, Gamse R: Substance P in peripheral sensory processes; in Poter R, O'Connor M (eds): Substance P in the Nervous System. Ciba Foundation Symposium 91. London, Pitman, 1982, pp 35–54.
Lembeck F, Holzer P: Substance P as a neurogenic mediator of antidromic vasodilation and plasma extravasation. Naunyn Schmiedebergs Arch Pharmacol 1979; 310:175–183.
Lembeck F, Starke K: Substanz P und Speichelsekretion. Naunyn Schmiedebergs Arch Exp Pathol Pharmakol 1968; 259:375–385.
Lundberg JM, Saria A: Capsaicin-induced desensitization of airway mucosa to cigarette smoke, mechanical and chemical irritants. Nature 1983;302:251–253.
Maggi A, Meli A: The sensory-efferent function of capsaicin-sensitive sensory neurons. Gen Pharmacol 1988;19: 1–43.
Nakanishi S: Mammalian tachykinin receptors. Annu Rev Neurosci 1991;14:123–136.
Nawa H, Hirose T, Takashima H, Inayama S, Nakanishi S: Nucleotide sequences of cloned cDNAs for two types of bovine brain substance P precursor. Nature 1983;306: 32–36.
Otsuka M, Konishi S, Takahashi T: A further study of the motoneuron-depolarizing peptide extracted from dorsal roots of bovine spinal nerves. Proc Jpn Acad 1972;48: 747–752.
Otsuka M, Yanagisawa M: Does substance P act as a pain transmitter. Trends Pharmacol Sci 1987;8:506–510.
Otsuka M, Yoshioka K: Neurotransmitter functions of mammalian tachykinins. Physiol Rev 1993;73:229–308.
Pernow B: Studies on substance P. Acta Physiol Scand 1953; 29(suppl 105).
Pernow B: Substance P. Pharmacol Rev 1983;35:85–141.
Piercey MF, Moon MW, Blinn JR: The role of substance P as a central nervous system neurotransmitter; in Jordan CC, Oehme P (eds): Substance P, Metabolism and Biological Actions. London, Taylor & Francis, 1985, pp 165–176.
Powell D, Leeman S, Tregear GW, Niall HD, Potts JT: Radioimmunoassay for substance P. Nat New Biol 1973; 241:252–254.

Rapport MM, Green AA, Page IH: Partial purification of the vasoconstrictor in beef serum. J Biol Chem 1948;174: 735.
Snider RM, Constantine JW, Lowe JA, Longo KP, Lebel WS, Woody HA, Drozda SE, Desai MC, Vinick FJ, Spencer RW, Hess HJ: A potent nonpeptide antagonist of substance P (NK_1) receptor. Science 1991;251:435–437.
Stoppini L, Barju F, Mathison R, Baertschi I: Spinal substance P transmits bradykinin but not osmotic stimuli from hepatic portal vein to hypothalamus in rat. Neuroscience 1984; 11:903–912.
Stricker S: Untersuchungen über die Gefässwurzeln des Ischiadicus. Sitzber Kais Akad Wiss Wien 1876;3:173–185.
Studer RO, Trzeciak A, Lergier W: Isolierung und Aminosäuresequenz von Substanz P aus Pferdedarm. Helv Chim Acta 1973;56:860–866.
Szallasi A, Blumberg PM: Vanilloid (capsaicin) receptors and mechanisms. Pharmacol Rev 1999;51:159–212.
Traurig H, Saria A, Lembeck F: The effects of neonatal capsaicin treatment on growth and subsequent reproductive function in the rat. Naunyn Schmiedebergs Arch Pharmacol 1984;327:254–259.
Tregear GW, Niall HD, Potts JD, Leeman SE, Chang MM: Synthesis of substance P. Nat New Biol 1971;232:87–89.
Ueda M, Kuraishi Y, Sugimoto K, Satoh M: Evidence that glutamate is released from capsaicin-sensitive primary afferent fibres in rats: study with on-line continuous monitoring of glutamate. Neurosci Res 1994;20:231–237.
von Euler US, Gaddum JH: An unidentified depressor substance in certain tissue extracts. J Physiol Lond 1931; 72:74–87.
von Euler US: Preparation of substance P. Scand Arch Physiol 1936;73:142–144.
von Euler US: Historical notes; in Von Euler US, Pernow P (eds): Substance P. Nobel Symposium 37. New York, Raven Press, 1977, pp 1–3.
Wimalawansa SJ: Calcitonin gene-related peptide and its receptors: Molecular genetics, physiology, pathophysiology, and therapeutic potentials. Endocr Rev 1996;17: 533–585.
Yamamoto A, Nakamura K, Furukawa K, Konishi Y, Ogino T, Higashiura K, Yago H, Okamoto K, Otsuka M: A new nonpeptide tachykinin NK1 receptor antagonist isolated from the plants of Compositae. Chem Pharm Bull (Tokyo) 2002;50:47–52.
Yanagisawa M, Murakoshi T, Tamai S, Otsuka M: Tail-pinch method in vitro and the effects of some antinociceptive compounds. Eur J Pharmacol 1984;106:231–239.
Zetler G, Schlosser L: Über die Verteilung von Substanz P und Cholinacetylase im Gehirn. Naunyn Schmiedebergs Arch Exp Pathol Pharmakol 1955;224:159–175.

Calcitonin Gene-Related Peptide

Whereas in the 1970s new neuropeptides were found mostly by extracting large amounts of brain or pituitary tissue and then monitoring the purification steps by bioassay, in the 1980s molecular cloning was introduced as a new method. A new peptide discovered in this way was calcitonin gene-related peptide (CGRP). CGRP was one of the first examples of a biologically active peptide to be discovered by the application of recombinant DNA technology in the absence of prior biological information.

The calcitonin gene was cloned first by Jacobs and collaborators in 1980, and further characterization was carried out subsequently [Jacobs et al., 1981]. The discovery of CGRP was made in 1983 by molecular cloning of the calcitonin gene by Rosenfeld et al. [1983]. They observed that serially transplanted tumours from a rat medullary thyroid carcinoma lost the ability to produce calcitonin. This spontaneous 'switch' from 'high' to 'low' or absent calcitonin production was accompanied by a slight increase in the size of the mRNA detected by calcitonin cDNA probes, which was later determined to be an 'altered mRNA' species derived from the calcitonin/CGRP gene. This altered mRNA was found to be derived from the same RNA precursor as a result of alternative RNA processing. The mechanism of regulation of calcitonin/CGRP production via alternative RNA splicing remains unclear.

The peptide deduced from the cDNA for the alternative product was named CGRP. Rosenfeld and coworkers deduced the amino acid sequence – CGRP is a 37-amino acid neuropeptide – they synthesized a peptide fragment and raised antibodies, which were then used for radioimmunoassay and immunohistochemistry. Immunoreactive CGRP was identified in a unique distribution in a large number of cell groups and pathways in the central nervous system distinct from any other known neuropeptide. They confirmed the production of CGRP mRNA in the brain and identified the sites of its biosynthesis. Tissue specificity of the RNA processing events was suggested because virtually no calcitonin mRNA could be identified in the rat brain. On the other hand, in the thyroid cells primarily calcitonin mRNA and its encoded peptide are present, beside a small amount of CGRP. Alternative splicing of the primary RNA transcript leads to the production of CGRP and calcitonin in a tissue-specific manner [Rosenfeld et al., 1992]. It has to be noted that it was later found that the 37-amino acid polypeptide sequence of CGRP can vary slightly among different mammalian species. Common to all is the amidated C-terminal.

CGRP turned out to be present in numerous neuronal systems, including neurons in the dorsal root ganglia, and in even higher concentrations and more frequently than the peptide par excellence in primary sensory neurons until then, substance P [Skofitsch and Jacobowitz, 1985a, b]. In fact, it could be established that virtually all substance P neurons contained CGRP [Hökfelt et al., 1992]. CGRP fibres form a dense network in lamina I and II of the dorsal horn, i.e. the central branches of thin unmyelinated primary sensory neurons. The CGRP fibres in the dorsal horn are depleted after capsaicin treatment, and this 'assay' has been used repeatedly to follow the efficacy of a 'capsaicin denervation' in rats [Holzer, 1991].

Of interest is also its coexistence with acetylcholine in motoneurons, although no specific function has been attributed to the peptide in motor neuron transmission [Van Rossum et al., 1997].

It can be stated that CGRP significantly stimulated the research on transmitter coexistence in neurons. Although it was known before that a peptide can coexist with a classical transmitter in one neuron, now it has become evident that several peptides can also coexist in one neuron. Classical transmitters are present in small, clear synaptic vesicles, and peptides in large dense core vesicles. Depending on the action potential firing rate, either one or both transmitters are released and exert their specific functions. This research area was very elegantly elucidated by Hökfelt et al. [1986] at the Karolinska Institute in Stockholm.

Studies with regard to the function of CGRP revealed that it has a strong vasodilator action and is an even more potent vasodilator than substance P [Brain and Williams, 1988]. Although CGRP does not cause plasma extravasation, by its vasodilator effect it can act in a proinflammatory manner, potentiating oedema formation and neutrophil accumulation induced by inflammatory mediators. It was soon recognized that CGRP – retrogradely released from primary sensory neurons (see also p. 181) – is actually the major mediator of neurogenic vasodilatation in the skin. Most probably by its vasodilator action CGRP plays a protective role regarding damage in the gastrointestinal mucosa [Holzer, 1994].

By the use of a peptidergic CGRP antagonist two receptors (CGRP-1 and CGRP-2 receptors) were postulated. The vasodilator reaction is mediated by CGRP-1 receptors, coupled to the adenylate cyclase system, and arises from a direct action on the vascular smooth muscle and does not involve nitric oxide or prostaglandins [Wimalawansa, 1996; Brain, 1997].

In several primary forms of headache such as migraine or cluster headache the trigeminal nerves can be activated by vasoactive components acting on the blood vessel wall. This initiates the trigeminovascular reflex, causing an antidromic release of CGRP [Moskowitz, 1993; Edvinsson, 2001]. CGRP is the major neuropeptide involved in this neurogenic vasodilator response within the dura mater and in the cerebral vasculature.

The activation of the trigeminovascular system with consequent neuropeptide release plays a crucial role in the pathophysiology of migraine. Whereas in animal models neurogenic substance P and CGRP release is relevant, in humans probably only CGRP release leads to a vasodilatation of cerebral arteries. CGRP is found elevated in the cranial venous blood of patients suffering acute attacks of migraine. Following administration of the 5-$HT_{1B/1D}$ receptor agonist sumatriptan, the CGRP levels returned to control. Triptans and also dihydroergotamine inhibit CGRP release by mediating hyperpolarization of sensory prejunctional nerve terminals following 5-$HT_{1B/1D}$ receptor activation [Limmroth et al., 2001]. By this way they limit overdistension of cerebral arterioles.

New non-peptide CGRP-1 receptor antagonists might be valuable tools for studying CGRP-mediated functions in migraine and neurogenic inflammation and for attenuating such symptoms. Any systemic application of CGRP antagonists, however, has to be seen in the light that CGRP is considered an endogenous cardioprotective substance in myocardial ischemia [Franco-Cereceda and Liska, 2000].

References

Brain SD: Sensory neuropeptides: their role in inflammation and wound healing. Immunopharmacology 1997;37: 133–152.

Brain SD, Williams TJ: Substance P regulates the vasodilator activity of calcitonin gene-related peptide. Nature 1988; 335:73–75.

Edvinsson L: Aspects of the pathophysiology of migraine and cluster headache. Pharmacol Toxicol 2001;89:65–73.

Franco-Cereceda A, Liska J: Potential of calcitonin gene-related peptide in coronary heart disease. Pharmacology 2000;60:1–8.

Hökfelt T, Arvidsson U, Ceccatelli S, et al: Calcitonin gene-related peptide in the brain, spinal cord, and some peripheral systems. Ann NY Acad Sci 1992;657:119–134.

Hökfelt T, Holets VR, Staines W, et al: Coexistence of neuronal messengers – an overview; in Hökfelt T, Fuxe K, Pernow B (eds): Coexistence of Neuronal Messengers: a New Principle in Chemical Transmission. Progress in Brain Research. Amsterdam, Elsevier Science, 1986, vol 68, pp 33–70.
Holzer P: Capsaicin: cellular targets, mechanisms of action, and selectivity for thin sensory neurons. Pharmacol Rev 1991;43:143–201.
Holzer P: Calcitonin gene-related peptide; in Walsh JH, Dockray GJ (eds): Gut Peptides: Biochemistry and Physiology. New York, Raven Press, 1994, pp 493–523.
Jacobs JW, Goodman RH, Chin WW, Dee PC, Habener JF, Bell NH, Potts JT: Calcitonin messenger RNA encodes multiple polypeptides in a single precursor. Science 1981; 213:457–459.
Limmroth V, Katsarava Z, Liedert B, Guehring H, Schmitz K, Diener HC, Michel MC: An in vivo rat model to study calcitonin gene-related peptide release following activation of the trigeminal vascular system. Pain 2001;92: 101–106.
Moskowitz MA: Neurogenic inflammation in the pathophysiology and treatment of migraine. Neurology 1993;43: S16–S20.
Rosenfeld MG, Emeson RB, Yeakley JM, Merillat N, Hedjran F, Lenz J, Delsert C: Calcitonin gene-related peptide: a neuropeptide generated as a consequence of tissue-specific, developmentally regulated alternative RNA processing events. Ann NY Acad Sci 1992;657:1–17.
Rosenfeld MG, Mermod JJ, Amara SG, Swanson LW, Sawchenko PE, Rivier J, Vale WW, Evans RM: Production of a novel neuropeptide encoded by the calcitonin gene via tissue specific RNA processing. Nature 1983;304: 129–135.
Skofitsch G, Jacobowitz DM: Calcitonin gene-related peptide: detailed immunohistochemical distribution in the central nervous system. Peptides 1985a;6:721–745.
Skofitsch G, Jacobowitz DM: Quantitative distribution of calcitonin gene-related peptide in the rat central nervous system. Peptides 1985b;6:1069–1073.
van Rossum D, Hanisch UK, Quirion R: Neuroanatomical localization, pharmacological characterization and functions of CGRP, related peptides and their receptors. Neurosci Behav Rev 1997;21:649–678.
Wimalawansa SJ: Calcitonin gene-related peptide and its receptors: molecular genetics, physiology, pathophysiology, and therapeutic potentials. Endocr Rev 1996;17: 533–585.

Neuropeptide Y, Galanin and Vasoactive Intestinal Polypeptide

A purely chemical strategy avoiding the use of biological assays was introduced by Victor Mutt from the Karolinska Institute in Stockholm and led to the discovery of several neuropeptides from extracts of gastrointestinal tissues in the 1970s and early 1980s. Determining chemical details characteristic of hormonal peptides or neuropeptides, e.g. a C-terminal α-amide structure, has led to the identification of neuropeptide tyrosine (NPY) and galanin. Vasoactive intestinal polypeptide (VIP) had already been discovered earlier.

Certain peptides do not contain a free carboxyl group at the C-terminus (i.e. COOH), but instead contain an amidated carboxyl group (i.e. $CONH_2$). During sequence studies on cholecystokinin (CCK), Mutt noted that the C-terminal phenylalanine amide residue could be isolated chromatographically. This observation led Tatemoto and Mutt [1978] to devise an assay system to detect, in crude tissue extracts, the presence of peptides containing such C-terminal α-amide structures. While monitoring the purification of CCK and secretin from porcine intestinal extracts with this assay system, they noted the presence of peptides containing C-terminal α-amide structures which could not be attributed to known peptides. They quickly recognized the significance of this finding and proceeded to isolate a series of previously unrecognized peptides, each of which has subsequently been shown to possess potent biological activities and to have a characteristic tissue distribution.

Vasoactive Intestinal Polypeptide

VIP, a polypeptide consisting of 28 amino acids, was discovered by Said and Mutt [1970; also 1972] in extracts of the porcine intestinal wall. First Said and Mutt detected the presence of a vasoactive peptide in lung tissue extracts, but since there was a far greater accessibility to intestinal extracts in Mutt's laboratory, they screened for and detected the presence of the presumed identical peptide in Mutt's intestinal extracts. Said and Mutt isolated an intestinal peptide based on its vasodilatory properties, which they named the vasoactive intestinal polypeptide or VIP; it has been assumed that the originally identified lung peptide is the same as that isolated from the intestine. On chemical characterization, VIP was found to be composed of 28-amino acid residues and to have an N-terminal His residue and a C-terminal Asn amide residue.

The first synthesis of the entire peptide was achieved by Bodanszky et al. [1973]. Originally described as an intestinal peptide hormone, it became soon evident that this peptide is also present in a variety of central nervous systems (CNS) and peripheral neurons. Actually, in the mammal VIP appears to be strictly confined to neural structures of the peripheral nervous system and CNS, hence VIP is a neuropeptide. An exception occurs in human pathology, where VIP is present in and released from certain islet cells or neural tumours known as VIPoma.

The whole range of the physiological functions of VIP is not clear yet; evidence points to its importance as a mediator or modulator of several basic functions [Said, 1984]. These functions include blood flow, smooth muscle relaxation, exocrine secretion, and endocrine/neuroendocrine activity. Many exocrine glands are supplied by networks of VIP-rich fibres. Some of these VIP-containing nerves are also cholinergic and co-release both acetylcholine (ACh) and VIP (see also below). A principal effect of VIP on exocrine glands is vasodilatation and increased blood flow; in addition VIP may stimulate salivary secretion in certain glands or potentiate the salivary secretory response to ACh and the pancreatic secretory response to CCK. In the lung VIP has been described as a possible mediator of airway smooth muscle relaxation, specifically of the so-called 'non-adrenergic, non-cholinergic (NANC) relaxation of airways'. VIP is known to stimulate the enzymatic breakdown of glycogen to glucose, probably by stimulating cyclic AMP production. It also exerts a glycogenolytic action in the nervous system. Furthermore, several modulations of sexual function and genital organs by VIP are known.

VIP as a potent vasodilator peptide co-exists with ACh in parasympathetic postganglionic nerves, and as elegantly shown for the first time by Lundberg et al. [1982] this co-transmission has significant functional effects. Their experiments in the cat salivary gland have shown that ACh upon its release through nerve activity evokes saliva secretion and an increase in blood flow, effects which are both atropine-sensitive. The released VIP is responsible for the vasodilatation in the gland but has no apparent effect on secretion. However, VIP potentiates ACh-induced secretion and when ACh and VIP are infused together additive effects on blood flow are observed.

VIP is also present in certain primary sensory neurons, but it is unlikely that it contributes to neurogenic vasodilatation or inflammation. An interesting phenomenon, which it shares with galanin, is the increase of VIP in the DRG and the superficial dorsal horn upon peripheral axotomy of the corresponding neurons (whereas substance P or CGRP decrease). The exact significance of this upregulation of VIP and galanin is not known [Shebab and Atkinson, 1986].

Two types of high-affinity VIP receptors, which have also been cloned, exist in mammalian tissues. One responds to VIP and to the structurally related peptide hormone pituitary adenylate cyclase-activating polypeptide (PACAP) and is named $VPAC_1$ receptor, and the other one is called $VPAC_2$ receptor [see Harmar et al., 2000].

Neuropeptide Y

NPY, a 36-amino acid neuropeptide, was discovered by Tatemoto et al. [1982]. It belongs to the pancreatic polypeptide family. The first member of this peptide group, avian pancreatic polypeptide (APP), was isolated from chicken pancreas as a by-product of insulin purification in the 1970s by Kimmel et al. [1975]. They noted the presence of a dominant chromatographic peak during the purification of avian insulin; this chromatographic peak contained a peptide they named the pancreatic polypeptide. Subsequently homologous peptides were found in the islets of Langerhans of all mammalian species. Since it was already known at that time that several gastrointestinal peptides also occur in the brain, it was logical to look for pancreatic polypeptide-like peptides in the mammalian CNS. Encouraging results were obtained using an antiserum directed against APP; however, the characterization of this peptide immunoreactivity turned out to be difficult. Despite the striking histochemical staining, which suggested an abundant CNS peptide, only trace amounts could be isolated and detected by radioimmunoassay. This problem was fortunately soon resolved when Tatemoto and Mutt reported in 1980 that porcine brain and gut contained large amounts of a pancreatic polypeptide-like peptide the amino acid sequence of which contained an amino-terminal tyrosine residue and a carboxy-terminal tyrosine amide [for a review, see Emson and De Quidt, 1985]. The peptide was named PYY (peptide YY, since it has tyrosine at both amino- and carboxy-terminals, Y being the abbreviation for tyrosine in the single letter amino acid code). The carboxy-terminal tyrosine amide was critical for its chemical identification (see above). PYY was thought to be present in both brain and gut, but subsequent work by Tatemoto et al. [1982] demonstrated that the brain pancreatic polypeptide-like peptide differed slightly from PYY and was thus called neuropeptide Y (NPY).

Following the availability of pig NPY for the production of antibodies and development of radioimmunoassays, several groups have reported the presence of NPY-like immunoreactivity in many regions of the mammalian nervous system [Lundberg et al., 1984]. It was soon evident that there was sufficient sequence homology between APP and NPY to allow APP-directed antisera to be used to visualize NPY histochemically, despite the absence of any significant cross-reactivity in radioimmunoassays – a quite unusual characteristic!

With the new knowledge and the new antisera, NPY-positive neurons were found throughout the CNS and peripheral nervous system. There exists a widespread distribution of NPY-containing cell bodies in the cerebral cortex including the hippocampal formation with a role in cardiovascular, neuroendocrine and vigilance control [Fuxe et al., 1990]. Of much interest is its potential role in the control of food intake and energy metabolism; in this respect NPY is one of the most powerful stimulants of food intake known today.

Most interestingly it was established that NPY co-exists with catecholamines in both central and peripheral neurons. In the periphery NPY co-exists with noradrenaline in sympathetic postganglionic neurons. Only a proportion of sympathetic noradrenergic nerves contain the NPY peptide and these neurons innervate blood vessels, the muscle of the heart, spleen and vas deferens. Nerve stimulation can cause a co-release of noradrenaline and NPY. Lundberg and Tatemoto [1982] have shown that NPY also causes vasoconstriction when released from sympathetic nerves. NPY exhibits a slowly developing, long-lasting effect whereas noradrenaline elicits the well-known short-lasting vasoconstrictor effect. However, other presynaptic modulatory effects by NPY have also been described in peripheral organs.

Based on the differential potency of NPY and certain of its C-terminal fragments, the NPY receptor system was originally classified into two subtypes, designated as Y1 and Y2. At present, five NPY receptor subtypes have been cloned and designated as Y_1, Y_2, Y_4, Y_5 and y_5, all belonging to the seven transmembrane G protein-coupled receptor family [Redrobe et al., 2002].

Galanin

Galanin, a 29-amino acid peptide, was first isolated by Tatemoto et al. [1983]. It was detected in and isolated from porcine upper small intestinal extracts using a chemical assay that detected the presence of a peptide containing the C-terminal α-alanine structure (see above). On sequence analysis, galanin was found to contain 29-amino acid residues and to have a unique amino acid sequence, an N-terminal residue of glycine, and a C-terminal alanine amide residue. Because it was isolated on a strictly chemical basis, galanin was named after its N- and C-terminal residues, glycine and alanine amide. Interestingly, as was found later, in humans galanin consists of 30 amino acids and is not amidated at the C-terminal.

Subsequent to its isolation, galanin was found to possess biological activity, exciting rodent smooth muscle and lowering plasma insulin levels [Crawley, 1995]. Numerous studies have demonstrated that galanin in the mammal is confined to neural structures, and the detailed distribution of galanin-like immunoreactivity in the CNS and peripheral nervous system has been reported.

Immunocytochemical mapping of galanin-like immunoreactivity in the brain revealed a wide distribution of galanin-containing neurons and receptors [Skofitsch and Jacobowitz, 1985]. Beside other locations galanin is present as a co-transmitter in cholinergic forebrain neurons. In intact primary sensory neurons galanin immunoreactivity can be seen only in a small number of neurons.

After peripheral axotomy, that is a lesion that can occur when the body is injured, there is a marked upregulation of galanin and also of VIP, whereas the prototypic neuropeptide transmitters like substance P and CGRP are markedly downregulated. The reasons for this 'paradox' upregulation of galanin and VIP are not quite understood. They could favour regeneratory responses or counteract pain responses (that can always occur following a peripheral axotomy as neuropathic pain), effects that have been demonstrated for galanin [Xu et al., 1990].

Three galanin receptor subtypes have been identified and cloned, GAL-R1, GAL-R2 and GAL-R3. Physiological and behavioural actions of galanin, as known today, include release of pituitary hormones, gastrointestinal and cardiovascular effects, upregulation following neural injury (see above), inhibitory effects on neuronal firing and neurotransmitter release, inhibitory effects on pain transmission, inhibitory effects on learning and memory and modulatory functions in seizures, depression and food intake [Crawley, 1999].

References

Bodanszky M, Klausner YS, Said SI: Biological activities of synthetic peptides corresponding to fragments and to the entire sequence of the vasoactive intestinal peptide. Proc Natl Acad Sci USA 1973;70:382–384.

Crawley JN: Biological actions of galanin. Regul Pept 1995; 59:1–16.

Crawley JN: The role of galanin in feeding behaviour. Neuropeptides 1999;33:369–375.

Emson PC, De Quidt ME: NPY – a new member of the pancreatic polypeptide family; in Bousfield D (ed): Neurotransmitters in Action. Amsterdam, Elsevier, 1985, pp 338–346.

Fuxe K, Agnati LF, Härfstrand A, et al: On the role of neuropeptide Y in information handling in the central nervous system in normal and physiopathological states. Ann NY Acad Sci 1990;579:28–67.

Harmar AJ, et al: International Union of Pharmacology. XVIII. Nomenclature of receptors for vasoactive intestinal polypeptide (VIP) and pituitary adenylate cyclase-activating polypeptide (PACAP). Pharmacol Rev 1998;50:265–270.

Kimmel JR, Hayden LJ, Pollock HG: Isolation and characterization of a new pancreatic polypeptide hormone. J Biol Chem 1975;250:9369–9376.

Lundberg JM, Änggård A, Fahrenkrug J, Lundgren C, Holmstedt B: Co-release of VIP and acetylcholine in relation to blood flow and salivary secretion in cat submandibular salivary gland. Acta Physiol Scand 1982;115:525–528.

Lundberg JM, Tatemoto K: Pancreatic polypeptide family (APP, BPP, NPY and PYY) in relation to sympathetic vasoconstriction resistant to alpha-adrenoceptor blockade. Acta Physiol Scand 1982;116:393–402.

Lundberg JM, Terenius L, Hökfelt T, Tatemoto K: Comparative immunohistochemical and biochemical analysis of pancreatic polypeptide-like peptides with special reference to presence of neuropeptide Y in central and peripheral neurons. J Neurosci 1984;4:2376–2386.

Redrobe JP, Dumont Y, Quirion R: Neuropeptide Y (NPY) and depression: from animal studies to the human condition. Life Sci 2002;71:2921–2937.

Said SI: Vasoactive intestinal polypeptide (VIP): current status. Peptides 1984;5:143–150.

Said SI, Mutt V: Polypeptide with broad biological activity: isolation from small intestine. Science 1970;169: 1217–1218.

Said SI, Mutt V: Isolation from porcine intestinal wall of a vasoactive octacosapeptide related to secretin and to glucagons. Eur J Biochem 1972;28:199–204.

Shebab SA, Atkinson ME: Vasoactive intestinal polypeptide (VIP) increases in the spinal cord after peripheral axotomy of the sciatic nerve originate from primary afferent neurons. Brain Res 1986;372:37–44.

Skofitsch G, Jacobowitz DM: Immunohistochemical mapping of galanin-like neurons in the rat central nervous system. Peptides 1985;6:509–546.

Tatemoto K, Carlquist M, Mutt V: Neuropeptide Y – a novel brain peptide with structural similarities to peptide YY and pancreatic polypeptide. Nature 1982;296:659–660.

Tatemoto K, Mutt V: Chemical determination of polypeptide hormones. Proc Natl Acad Sci USA 1978;75:4115–4119.

Tatemoto K, Rökaeus A, Jörnvall H, McDonald TJ, Mutt V: Galanin – a novel biologically active peptide from porcine intestine. FEBS Lett 1983;164:124–128.

Xu XJ, Wiesenfeld-Hallin Z, Villar MJ, Fahrenkrug J, Hökfelt T: On the role of galanin, substance P and other neuropeptides in primary sensory neurons of the rat: studies on spinal reflex excitability and peripheral axotomy. Eur J Neurosci 1990;2:733–743.

Opioid Peptides

Eight years before the discovery of the first opioid peptides in the brain, researchers had concluded that the complex interactions among the exogenous morphine-like drugs, antagonists, and what were then called mixed agonist-antagonists can only be explained by postulating the existence of more than one type of opioid receptor [Martin, 1967]. In 1973, following a methodological approach developed by Avram Goldstein and coworkers [see Goldstein, 1976], three groups of investigators [Pert and Snider, 1973; Simon et al., 1973; Terenius, 1973] independently reported the discovery of saturable, stereospecific binding sites or receptors for opioid drugs in the mammalian nervous system. Shortly thereafter, in 1975, Hughes et al. [1975] described the isolation from pig brain of two pentapeptides that exhibited morphine-like actions on the guinea pig ileum – actions that were specifically antagonized by naloxone. Within the same year, Goldstein and colleagues reported the presence of peptide-like substances in the bovine pituitary gland with opioid activity. This substance proved to be a polypeptide with 31-amino acid residues; it, too, exhibited opioid-like actions that were antagonized by naloxone. Hughes and Kosterlitz named the pentapeptides leucine- and methionine-enkephalin. The larger peptide was designated β-endorphin.

Subsequent research revealed that there are three distinct families of opioid peptides: the enkephalins (leucine-enkephalin and methionine-enkephalin), the endorphins (β-endorphin), and the dynorphins. Each family is derived from a genetically distinct precursor polypeptide and has a characteristic anatomical distribution [Höllt, 1985].

The enkephalins and dynorphins are contained primarily in interneurons with short axons in areas of the central nervous system (CNS) related to the perception of pain. Upon their release they appear to function as neurotransmitters or modulators of neurotransmission [Kosterlitz, 1980]. An important location for β-endorphin is the pituitary and it is assumed to function as a neurohormone.

Shortly after the enkephalins were discovered it was noticed that they were almost immediately hydrolyzed into their constituent amino acids when brought into contact with brain tissues either in vivo by local injection or in vitro with tissue extracts [Chruściel, 1980]. They were metabolized within a few minutes. Their binding characteristics could only be determined with accuracy in the presence of opiate inhibitors. β-Endorphin is more stable and is released under conditions of severe stress from the pituitary into the bloodstream.

The discovery of endogenous ligands with high affinities for opioid receptors has added new dimensions to the concept of the diversity of the opioid receptors. The enkephalins shared the pharmacological properties of the agonistic opioid alkaloids like morphine.

Long before the opiate receptors were cloned and their structure elucidated, a wide body of biochemical, pharmacological and behavioural evidence indicated that opiates exert their wide spectrum of actions by interaction with μ-, δ- and κ-opiate receptors [Jaffe and Martin, 1985]. The μ-receptor was operationally defined as the high-affinity site at which morphine-like opiates produce analgesia and a variety of other classical opiate effects [Terenius, 1978]. The δ-receptor was defined pharmacologically as the receptor which is found in peripheral tissues such as the mouse vas deferens, as well as in the CNS, and which exhibits a higher affinity for the naturally occurring enkephalins. The κ-receptor was that at which ketocyclazocine-like opiates produce analgesia, as well as their unique ataxic and sedative effects. Later this receptor has been defined as the receptor highly selective for dynorphin. β-Endorphin acts equally at the μ- and the δ-receptor [Simon and Hiller, 1978].

As already shortly outlined above, the decisive step forward in the research of the profile of opiate actions, of their receptors and of discovery of the endogenous opioid peptides came from the work of Hans W. Kosterlitz in Aberdeen. To motivate young scientists who have not yet achieved their decisive breakthrough in young years it is worth mentioning that Kosterlitz had made his famous discoveries in the Unit for Research on Addictive Drugs established for him to continue his research after retiring (!) from the chair of Pharmacology in the University of Aberdeen. He described his reminiscences of the important discoveries in a very personally written review [Kosterlitz, 1979]. Kosterlitz had established the simple and most effective bioassays to test opioid drugs: one assay was based on the electrical field stimulation of strips consisting of the myenteric plexus and the longitudinal muscle of the guinea pig ileum. As was known from the inhibitory action of morphine on intestinal peristalsis, this simple model seemed useful to analyze the mode of action of morphine; morphine inhibited the neurally evoked contractions. From investigations of the release of the transmitter acetylcholine, it appeared that the effect of opiates was presynaptic and that there were distinct receptors. Now it was possible to express the action of opiates in quantitative terms, and to assay agonist and antagonist action.

Another assay used was the mouse vas deferens stimulated via the hypogastric nerve; this neuronal transmission was depressed by morphine due to a reduction in the release of noradrenaline from the nerve terminals. There are differences in the pharmacological responses of these two bioassays to pure opiate agonists, to compounds with dual agonist and antagonist actions, and to pure antagonists. The concept of different opioid receptors was developed and Kosterlitz and his coworkers became more and more convinced that these receptors were not only designed to interact with the alkaloids from the poppy plant. The almost simultaneous findings of E.J. Simon in New York, S.H. Snider in Baltimore, and L. Terenius in Uppsala that membrane fragments of brain tissue had specific binding sites for opioids led to the search for compounds in the CNS that interacted specifically with the opioid receptors.

Kosterlitz joined forces with John Hughes and the latter undertook the mission of extracting potential opioid ligands from pig brain tissues obtained from the slaughterhouse. The bioassays soon indicated beyond doubt the presence of opioid compounds; however, the identification proved difficult. It became clear that they were not alkaloid compounds but peptides; yet the isolation of sufficient material and the identification of the structure of the small peptides took 2 years. In this effort they were first misled by the assumption that they were dealing with only one compound. The sequence of the first four amino acids was clear with Tyr-Gly-Gly-Phe, but the allocation of the methionine and leucine, which were shown to be present by amino acid analysis, was difficult. Finally Howard Morris of the Imperial College of Science and Technology in London established the presence of the two pentapeptides, methionine-enkephalin and leucine-enkephalin, by mass spectrometry.

Basically at the same time the isolation and sequencing of a larger opioid peptide later named β-endorphin was achieved by H. Li, A. Goldstein and R. Guillemin (San Francisco/La Jolla, Calif.) [Goldstein, 1976]. Thus the story of the opioid peptides started as outlined by Hans Kosterlitz.

Since the biological half-life of the two naturally occurring enkephalins is very short, many stable analogues with a high affinity to the opiate receptors and strong analgesic activity have been synthesized [Jaffe and Martin, 1985]. Immunocytochemical studies have shown the existence of separate enkephalin-, of β-endorphin- and later of dynorphin-containing areas in the brain [Bloom, 1983; Akil et al., 1984]. Whereas the β-endorphin system is limited to a few areas in the midbrain and limbic structures, as well as the pituitary gland, enkephalin-containing cell bodies and nerve processes are widely distributed throughout the brain and spinal cord. The enkephalins appear to be particularly concentrated in brain areas concerned with sensory transmission, endocrine control, respiration, motor activity, and behaviour. There is also a rich enkephalinergic innervation of the gastrointestinal tract. The distribution of opiate receptor binding sites and enkephalin-immunoreactive fibres shows a close parallelism.

Following the discovery of the opioid peptides there were questions about their possible physiological function. Certainly we are not dependent on the opioid peptides as shown by the lack of a withdrawal syndrome after the injection of the antagonist naloxone. On the other hand, it has been shown that during pathophysiological conditions such as stress, long-lasting physical exercise, or inflammation increased amounts of opioid peptides are probably released and can limit the experience of pain or induce euphoric changes of mood. The effect of β-endorphin is long-lasting as it is more enzyme-resistant, in contrast to the transient effects of the enkephalins. The enkephalins are good candidates for a possible role of inhibitory transmitters or neuromodulators [Beaumont and Hughes, 1979]. An interesting new finding in the 1980s was the presence of opioid peptides in inflammatory and immune cells in peripheral tissues and the peripheral analgesic action of endogenous opioid peptides [see Stein, 1995].

The cloning of the opiate receptors many years later and the availability of opiate receptor knockout mice did not really change the field of opioid peptides and opioid analgesics anymore. Three different G protein-coupled transmembrane receptors, each possessing the properties of the μ-, δ-, or κ-receptors, respectively, have been cloned. When expressed they showed the anticipated selectivity in both binding and functional assays for their respective ligands [Reisine, 1993; Uhl et al., 1994].

Transgenic mice with a lack of either the β-endorphin or the enkephalin gene did show some altered behaviour in response to noise or movement or altered responses to pain stimuli; however, it became clear that the recruitment of compensatory mechanisms – opioid or non-opioid – can keep an inhibitory system active [Mogil and Grisel, 1998]. By deleting the gene for either the μ-, δ-, or κ-opioid receptor specific knockout mice became available. Clearly when one opioid receptor is not present in an animal, the specific opioid agonists are devoid of activity (e.g. they display no more analgesic activity). There was only a marginal change in basal nociceptive behaviour. The experiments with opioid receptor knockout mice also made clear that the endogenous opioid peptides like the enkephalins are not totally receptor-specific, but exert their effects by interacting with the μ- and the δ-receptor. This was already known from pharmacological experiments [Mogil and Grisel, 1998].

These new methods confirmed what was already known and show that is not possible, at least at present, to separate the useful pharmacological effects such as analgesia from the unwanted side effects such as obstipation or respiratory depression, since the receptors responsible for these effects have a wide distribution and no specific receptor subtypes have been found.

References

Akil H, Watson SJ, Young E, Lewis ME, Khatchaturian H, Walker JM: Endogenous opioids: biology and function. Annu Rev Neurosci 1984;7:223–255.

Beaumont A, Hughes J: Biology of opioid peptides. Annu Rev Pharmacol Toxicol 1979;19:245–267.

Bloom FE: The endorphins: a growing family of pharmacologically pertinent peptides. Annu Rev Pharmacol Toxicol 1983;33:397–415.

Chruściel TL: Metabolism of enkephalins and endorphins; in Marsan CA, Traczyk WZ (eds): Neuropeptides and Neural Transmission. New York, Raven Press, 1980, pp 209–216.

Goldstein AG: Opioid peptides (endorphins) in pituitary and brain. Science 1976;193:1081–1086.

Höllt V: Multiple endogenous opioid peptides; in Bousfield D (ed): Neurotransmitters in Action. Amsterdam, Elsevier, 1985, pp 188–193.

Hughes J, Smith TW, Kosterlitz H, Fothergrill LA, Morgan BA, Morris HR: Identification of two related pentapeptides from the brain with potent opiate agonist activity. Nature 1975;258:577–579.

Jaffe JH, Martin WR: Opioid analgesics and antagonists; in Goodman A, Goodman LS, Rall TW, Murad F (eds): The Pharmacological Basis of Therapeutics, ed 7. New York, Macmillan, 1985, pp 491–531.

Kosterlitz HW: The best laid schemes o'mice an' men gang aft agley. Annu Rev Pharmacol Toxicol 1979;19:1–12.

Kosterlitz H: Enkephalins, endorphins, and their receptors; in Marsan CA, Traczyk WZ (eds): Neuropeptides and Neural Transmission. New York, Raven Press, 1980, pp 191–197.

Martin WR: Opioid antagonists. Pharmacol Rev 1967;19: 463–521.

Mogil JS, Grisel JE: Transgenic studies of pain. Pain 1998; 77:107–128.

Pert CB, Snyder SH: Opiate receptor: demonstration in nervous tissue. Science 1973;179:1011–1014.

Reisine T, Bell GI: Molecular biology of opioid receptors. Trends Neurosci 1993;16:506–510.

Simon EJ, Hiller JM: The opiate receptors. Annu Rev Pharmacol Toxicol 1978;18:371–394.

Simon EJ, Hiller JM, Edelman I: Stereospecific binding of the potent narcotic analgesic ^{3}H-etorphine to rat brain homogenate. Proc Natl Acad Sci USA 1973;70:1947–1949.

Stein C: The control of pain in peripheral tissue by opioids. N Engl J Med 1995;332:1685–1690.

Terenius L: Stereospecific interaction between narcotic analgesics and a synaptic plasma membrane fraction of rat cerebral cortex. Acta Pharmacol Toxicol 1973;32: 317–320.

Terenius L: Endogenous peptides and analgesia. Annu Rev Pharmacol Toxicol 1978;18:189–204.

Uhl GR, Childers S, Pasternak G: An opiate-receptor gene family reunion. Trends Neurosci 1994;17:89–93.

Nitric Oxide

Nitric oxide (NO), an unstable gas, has been used for many years in cases of angina pectoris and other blood supply problems without realizing that NO might actually be the endogenous transmitter. Drugs that release NO, such as nitroprusside and nitroglycerines, have been successfully used for decades. Alfred Nobel, who invented nitroglycerine, was prescribed the drug himself to support his ailing heart. Furchgott and Zawadzki [1980] were the first to deduce that a mysterious transmitter that is responsible for blood vessel relaxation is NO. Moncada et al. [1991] provided further evidence that the vasodilator action of several transmitters actually depends on the formation of NO.

It was found that the metabolism that turns arginine into citrulline releases NO and that NO might play different roles in cell-to-cell communication [Moncada et al., 1989]. Actually, within a short space of time, the most unlikely of molecules has become recognized as a biological mediator of importance to the function of almost every bodily organ.

The existence of NO in the brain was first suggested by demonstrations that cerebellar neuronal cultures release a factor with properties resembling NO, as well as by observations of NO-forming activity in brain extracts and slices. Soon after these early discoveries, NO synthase was discovered in neurons, and theories were developed as to what functions NO might have [Snyder, 1992a, b]. Arguably, the strangest place to find NO was in the brain itself, where amino acids had previously been thought to be the simplest type of chemical used for signalling between cells, but which are now known to have a rich capacity for making NO.

The special properties of NO set it apart from any other type of signalling system, the most important distinction being the ability of NO to diffuse freely and rapidly across membranes so that, once formed, NO is able to spread in three dimensions to act on neighbouring cellular elements, irrespective of anatomical connectivity. In this regard, NO provides an exception to the rule, formulated over a century ago, whereby the transfer of information between neurons occurs unidirectionally and at anatomically defined sites (synapses). This unique feature, of course, raises a number of problems, not least of which is how any sort of specificity can be incorporated into a spatially diffuse signal like NO [Kiss and Vizi, 2001; Paton et al., 2002].

The enzyme that generates NO, nitric oxide synthase (NOS), has a very wide distribution in the brain, comparable to the distribution of the major neurotransmitters glutamate and GABA, suggesting an involvement in a large number of different brain functions [Garthwaite, 1995, 2000]. Neuronal NOS is now termed nNOS; it was also localized in peripheral nerves and in non-nervous tissue. The other two forms of NOS characterized are called eNOS (from endothelial) und iNOS (from inducible). The pattern of staining of nNOS immunoreactivity is not restricted to synaptic vesicles or to the plasma membrane, but appears to be largely cytoplasmic throughout the nerve cell body, axon and varicose fibres [Martin, 2000]. Such a pattern suggests the possibility of release of the transmitter along the entire length of the nerve fibre rather than exclusively at varicose regions, as with all other transmitters. Equally widely distributed, and in a correlated fashion, is the accumulation of guanosine 3′,5′-monophosphate (cGMP) in response to NO. cGMP stimulates protein phosphorylation. This further reinforces the idea that the principal physiological 'receptor' for NO is the soluble guanylate cyclase. NO and cGMP are thus likely to be functional partners throughout the central nervous system (CNS).

The role of NO in the peripheral nervous system is connected with non-adrenergic, noncholinergic (NANC) nerves. First in the periphery NOS could be found in the endothelial lining of blood vessels. However, in cerebral arteries and in arteries to the penis, NOS is localized to the nerve plexus in the outer, adventitial layer of vessels. In these blood vessels NO is a neurotransmitter for vascular relaxation, the so-called NANC relaxation (for other mediators of an NANC relaxation of smooth muscle, see also the following chapter on ATP and Adenosine) [Martin, 2000]. nNOS also occurs in the myenteric plexus of neurons and mediates the physiological relaxation of the gut.

Beside its functions in the peripheral nervous system the functions ascribed to NO in the CNS range from synaptic plasticity in several of its forms, e.g. long-term potentiation, depression of synaptic transmission, synaptogenesis and hyperalgesia, to the modulation of sensory and motor pathways, the local regulation of blood flow and other functions [Hölscher, 1997; Paton et al., 2002].

Overall it turned out that NO is involved in what is called non-synaptic communication between neurons. This communication without synaptic connections and without the requirement of specific receptors is also of physiological importance. It has been demonstrated for NO in the interplay of glutamatergic and monoaminergic transmission strength. NO can be regarded now as an endogenous neurotransmitter, neuromodulator and intercellular messenger [Garthwaite, 2000; Kiss and Vizi, 2001].

References

Furchgott RF, Zawadzki JV: The obligatory role of endothelial cells in the relaxation of arterial smooth muscle by acetylcholine. Nature 1980;288:373–376.

Garthwaite J: Neural nitric oxide signalling. Trends Neurosci 1995;18:51–52.

Garthwaite J: The physiological roles of nitric oxide in the central nervous system; in Mayer B (ed): Handbook of Experimental Pharmacology. Berlin, Springer, 2000, vol 143, pp 259–275.

Hölscher C: Nitric oxide, the enigmatic neuronal messenger: its role in synaptic plasticity. Trends Neurosci 1997;20: 298–303.

Kiss JP, Vizi ES: Nitric oxide: a novel link between synaptic and nonsynaptic transmission. Trends Neurosci 2001; 24:211–215.

Martin W: The role of nitric oxide in the peripheral nervous system; in Mayer B (ed): Handbook of Experimental Pharmacology. Berlin, Springer, 2000, vol 143, pp 277–313.

Moncada S, Palmer RM, Higgs EA: Biosynthesis of nitric oxide from *L*-arginine. A pathway for the regulation of cell function and communication. Biochem Pharmacol 1989; 38:1709–1715.

Moncada S, Palmer RM, Higgs EA: Nitric oxide: physiology, pathophysiology, and pharmacology. Pharmacol Rev 1991;43:109–142.

Paton JFR, Kasparaov S, Paterson DJ: Nitric oxide and autonomic control of heart rate: a question of specificity. Trends Neurosci 2002;25:626–631.

Snyder SH: Nitric oxide and neurons. Curr Opin Neurobiol 1992a;2:323–327.

Snyder SH: Nitric oxide: first in a new class of neurotransmitters. Science 1992b;257:494–496.

ATP and Adenosine

Interest in ATP (adenosine 5′-triphosphate) as a fast synaptic transmitter dates back to the early 1960s. The application of intracellular microelectrode techniques by Geoffrey Burnstock and Mollie Holman [1961] led to the discovery of excitatory junction potentials in smooth muscle cells of the vas deferens. These rapidly rising depolarizations that are elicited by stimulation of the sympathetic nerve supply to the organ were not abolished by blockers of acetylcholine receptors and adrenoreceptors. They were also not mimicked by noradrenaline but rather by ATP, and they were not blocked by an ACh receptor antagonist. Later Burnstock investigated inhibitory nerves in the gut. He observed that transmural stimulation of the guinea pig taenia coli with single impulses of short duration produced large hyperpolarizations or inhibitory junction potentials in smooth muscle cells, which were associated with relaxation, and which persisted in the presence of adrenergic and cholinergic blockers. At the same time other investigators described that some of the inhibitory fibres to the vertebrate stomach were not adrenergic. When sufficient evidence was accumulated that the principal substance released from these non-adrenergic and noncholinergic (NANC) nerves is a purine nucleotide, Burnstock [1971] proposed that these nerves should be called 'purinergic' and that ATP was their transmitter substance. Although the assumption to see ATP as a neural transmitter was initially disparaged, it is now generally accepted. Originally 'purinergic' nerves were defined as an addition to 'adrenergic' and 'cholinergic' autonomic nerves. Today it is clear that besides ATP also several other transmitter compounds like peptides can be regarded as NANC transmitters in the autonomic nervous system (see, for example, sections on VIP, NPY and nitric oxide, pp. 193–196 and 201/202).

Burnstock [1972] studied the NANC relaxations in the gut and provided progressively sufficient – mainly indirect – evidence that adenosine triphosphate or a related nucleotide is the NANC transmitter released by gut nerves:

(1) ATP and the enzymes necessary for its formation are present in NANC nerves. Tritiated adenosine is taken up by the nerves, transformed into, and stored as ATP in such a way that it is available for release during nerve stimulation.

(2) ATP and its breakdown products [AMP (adenosine 5′-monophosphate), adenosine and inosine] are released into the perfusate during stimulation of NANC inhibitory nerves.

(3) The response of smooth muscle to ATP closely mimics the response to nerve stimulation. ATP and ADP (adenosine 5′-diphosphate) are the most active adenyl compounds, AMP and adensoine are about 100 times less active.

(4) Enzymes capable of breaking down ATP to less active derivatives are present in tissues innervated by NANC nerves.

(5) When tachyphylaxis is induced by repeated administration of ATP, the inhibitory responses to non-adrenergic nerve stimulation are abolished, while the inhibitory responses to adrenergic nerve stimulation persist.

Burnstock also characterized the structure of purinergic nerves in the Auerbach plexus of the gut and elucidated the electrophysiology of purinergic transmission as inhibitory junction potentials in the smooth muscles of the gut.

Others have characterized an ATP-mediated excitatory transmission in the vasa deferentia of various rodent species, where ATP is a co-transmitter of noradrenaline and NPY [Stjärne and Lundberg, 1986]. The excitatory junction potentials in the vas deferens smooth muscle cells or the first and fast phase of the twitch contractile response to nerve stimulation are mediated by ATP, whereas noradrenaline mediates a second and sustained tonic contraction. In the meantime ATP is regarded as a general constituent of all noradrenaline storage vesicles in sympathetic nerves. Besides the postsynaptic action of ATP on purinergic receptors in smooth muscle cells, it may also inhibit the release of noradrenaline and ATP via prejunctional purinoreceptors. In parasympathetic nerves, ATP might act as a co-transmitter with acetylcholine [Burnstock, 1986].

A specific transmitter role for adenosine by itself has also been established in the meantime, specifically in the spinal cord and in the periphery, involved in the modulation of pain signalling during inflammation and nerve injury [Sawynok and Liu, 2003].

ATP acts at cell surface receptors of two fundamentally distinct types – ligand-gated ion channels (P2X receptors) and G protein-coupled receptors (P2Y receptors) [Zimmermann, 1994; Surprenant et al., 1995]. Adenosine can exert its actions by activation of a separate class of G protein-coupled cell surface receptors, the adenosine A1, A2A, A2B and A3 receptors. The P2X receptors have a function in smooth muscle contraction in bladder and vas deferens and in nociception, the P2Y receptors in vascular relaxation; adenosine is an important neuromodulator involved in pain transmission in the spinal cord and the periphery.

References

Burnstock G: Neural nomenclature. Nature 1971;229:282–283.

Burnstock G: Purinergic nerves. Pharmacol Rev 1972;24:509–581.

Burnstock G: Purines as cotransmitters in adrenergic and cholinergic neurons; in Hökfelt T, Fuxe K, Pernow B (eds): Coexistence of Neuronal Messengers: a New Principle in Chemical Transmission. Progress in Brain Research. Amsterdam, Elsevier Science, 1986, vol 68, pp 193–203.

Burnstock G, Holman ME: The transmission of excitation from autonomic nerve to smooth muscle. J Physiol 1961;155:115–133.

Sawynok J, Liu XJ: Adenosine in the spinal cord and periphery: release and regulation of pain. Prog Neurobiol 2003;69:313–340.

Stjärne L, Lundberg JM: On the possible roles of noradrenaline, adenosine 5′-triphosphate and neuropeptide Y as sympathetic cotransmitters in the mouse vas deferens; in Hökfelt T, Fuxe K, Pernow B (eds): Coexistence of Neuronal Messengers: a New Principle in Chemical Transmission. Prog Brain Res. Amsterdam, Elsevier Science, 1986, vol 68, pp 263–278.

Surprenant A, Buell G, North RA: P_{2X} receptors bring new structure to ligand-gated ion channels. Trends Neurosci 1995;18:224–229.

Zimmermann H: Signalling via ATP in the nervous system. Trends Neurosci 1994;17:420–426.

Peripheral Neurogenic Stimulators

In this chapter, we focus on transmitters and functions of afferent nerves, particularly the non-myelinated nerves subserving nociceptive and other functions. The peripheral nervous system consists of the following principal elements: (1) somatic and visceral afferent nerves (sensory neurons), (2) somatic efferent nerves, innervating skeletal muscles (motor neurons), and (3) autonomic efferent nerves including the enteric nervous system.

'Orders' and 'Messages'

Motor neurons consist of a cell body with a nucleus, the metabolic centre, dendrites receiving signals from other neurons, and the axon with its peripheral terminal. For a motor neuron supplying a muscle in the foot, it has been calculated that if the size of the cell body were the size of a tennis ball, the dendrites of the cell would fill an average-sized living room and the axon would be 1.6 km long, although only 13 mm in diameter [Ganong, 1991].

The axon serves the propagation of an electrical potential in an all-or-none way, called the action potential. The conductance is an active, self-propagating process. The mechanism of conductance is based on the exchange of ions via various types of ion channels, of which there are 50–1,200/μm^2 in the membrane.

Efferent fibres conduct 'orders' to the periphery. The neurotransmitter, synthesized in the neuron, is released from the peripheral terminal and activates the appropriate receptor located opposite the nerve terminal. Efferent nerves of the autonomic nervous system act on smooth muscles in blood vessels and viscera, and on secretory organs.

The cell body of afferent fibres is localized outside the spinal cord in the dorsal root ganglia. The neuron is pseudo-unipolar. The peripheral fibre, which might be much longer than the central dorsal root fibres, ends in peripheral tissues. All afferent neurons transmit 'messages' to the central nervous system.

The central terminals of afferent fibres have synaptic contact with second-order neurons in the gray matter of the dorsal horn of the spinal cord or in the brainstem. These dorsal horn neurons project axons and transmit messages to higher brain centres, including the reticular formation and the thalamus. The third-order afferent neuron ultimately projects to the cerebral cortex. Not surprisingly, the neural circuitry in the dorsal horn is already incredibly complex.

Dorsal root ganglion cells are classified into (1) proprioceptive neurons from specialized structures in muscle cells sensing muscle stretch, (2) low threshold mechanosensitive neurons transmitting mechanical stimuli from specific assemblies, and (3) temperature- and pain-sensing neurons that terminate as free nerve endings in the skin, around blood vessels and hair follicles.

Aδ and C Fibres

Small- and medium-diameter cell bodies give rise to slowly conducting, thinly myelinated Aδ or unmyelinated C fibres. Aδ fibres respond to mechanical stimuli, but are also affected by intense heat or tissue injury. Most C fibres respond to thermal and mechanical stimuli. Others are mechanically insensitive, but respond to noxious heat. Most C fibres also respond to noxious chemical stimuli, such as acid or capsaicin (see below), and are therefore 'polymodal'. As they transmit pain, they are called nociceptors.

Aδ and C fibres should be given special attention as they have two different functions:

(1) Stimulation of the peripheral terminal induces the propagation of an impulse to the central terminal, where it transmits the information to a second-order afferent neuron.

(2) In addition, the stimulation of nociceptive neurons causes the release of transmitters such as substance P and CGRP from the peripheral terminals, which induce local vasodilatation and plasma protein extravasation, seen as the 'wheel and flare' reaction in the skin or also called the 'axon reflex' response.

Neurotransmitters Released upon Antidromic Nerve Stimulation

Experimentally, stimulation of afferent neurons can be achieved electrically by the 'antidromic nerve stimulation', after cutting the nerve centrally to the site of stimulation. The induced action potentials can propagate only in the antidromic direction to the peripheral terminal and this causes a vasodilatation called the 'antidromic vasodilatation'.

This peripheral effect has a long history, beginning with its first observation by Stricker [1876] and its confirmation in the dog by Bayliss [1901]. Hinsey and Gasser [1930] showed that the antidromic vasodilatation needs a stimulus strong enough to stimulate nociceptors. Celander and Folkow [1953] found that only pain fibres induce the 'axon reflex'. Lembeck and Holzer [1979] also demonstrated the existence of the 'antidromic vasodilatation' and of the 'neurogenic inflammation', i.e. plasma protein extravasation (see below), in the rat.

Definitively, the release of neurotransmitters from sensory neurons must be responsible for the antidromic vasodilatation. Dale [1935] suggested that the knowledge of the substance which causes the antidromic vasodilatation might hint at the nature of the function of the central terminal of sensory afferents.

Pharmacological experiments led to the discovery of the neurotransmitter substance P in afferent nerves [Lembeck, 1953] (see also pp. 177–189). The antidromic vasodilatation, and also the evoked plasma protein extravasation, could be mimicked by intra-arterial injection of substance P [Lembeck and Holzer, 1979].

The vasodilatation evoked by antidromic nerve stimulation for 1 min lasted for more than 10 min. Gamse et al. [1980] showed that the plasma protein extravasation results from the release of histamine from mast cells by substance P. CGRP is responsible for the vasodilatation only. The long duration, exceeding the time of stimulation, might result from the formation of bradykinin and subsequently a long-lasting release of prostaglandins from the surrounding tissue [Lembeck et al., 1976; Guo et al.,1999]. As these effects are similar to inflammatory responses, they have already previously been called 'neurogenic inflammation' by Jancsó et al. [1967].

The 'Nocifensor System'

The neurogenic inflammation, evoked by antidromic stimulation, might have a physiological significance: vasodilatation might play a role in the removal of intrinsic or extrinsic toxic material by an increase in blood flow. Plasma protein extravasation might have the same function by enhanced lymphatic drainage.

This is reminiscent of the work of Sir Thomas Lewis in 1937, which was only based on careful observation of the human skin as regards wheel and flare reaction and hyperalgesia around a skin injury. He noted that the reactions are mediated by the dorsal roots but are independent of an activity arising in the central nervous system or in sympathetic nerves. Lewis regarded it as the first line of defence, calling it the 'nocifensor system' [for a review, see Lembeck, 1983].

Transmitters at the Central Terminals of Afferent Neurons

It was also shown that substance P was released from central terminals of afferent fibres in experiments in vivo and on spinal cord slices in vitro. These findings indicated that substance P could be the transmitter at the central endings [Gamse et al., 1979; for a review, see Holzer, 1991]. The disappointment came with the discovery of the non-peptide substance P antagonists, which completely and specifically blocked the actions of substance P at the peripheral terminals, but not at the central terminals [Lembeck et al., 1992].

At that time, De Biasi and Rustioni [1988] discovered the co-existence of glutamate and substance P. Ueda et al. [1994] developed an online monitoring method, by which the release of glutamate, along with substance P, could be shown from spinal cord slices. This means that glutamate is released from central terminals of sensory afferents. The released glutamate acts on NMDA receptors. Reflex reactions which were evoked by afferent nerve stimulation were completely eliminated by the NMDA receptor antagonist MK-801 [Donnerer and Amann, 1994], i.e. the reflexes were mediated by glutamate. In spinal cord slices, the release of glutamate could be evoked by capsaicin and by substance P. This release was not influenced by MK-801, but by a substance antagonist. This means that glutamate is the essential transmitter at central terminals, while substance P might only have an auxiliary function [Juránek and Lembeck, 1996, 1997]. On the other hand, the plasma protein extravasation induced by antidromic nerve stimulation remained fully intact with MK-801 [Lembeck and Sieler, 1999]. This means that the transmitter at peripheral terminals is only peptidergic (substance P and CGRP).

The Axon

The axon plays a role in the conduction or propagation of a transient electrical potential in an all-or-none way called the action potential. Action potentials are generated in the nerve terminals of afferent fibres. The conductance is an active, self-propagating process. The impulse moves along the nerve fibre at a constant amplitude and velocity without any loss.

The velocity of conductance is related to the diameter of the nerve fibre. Afferent A fibres with a diameter of 12–20 μm conduct at a speed of 70–120 m/s (comparable to the top speed of a racing car), the C fibres with a diameter of 0.3–1.3 μm at a speed of 0.7–2.3 m/s (comparable to the speed of the racing champion when jogging). The magnitude of the stimulation, for example by heat, is converted into a frequency code.

The axon also transports substances (for example, substance P synthesized in the cell body) to the terminals by either a fast (400 mm/day) or a slow (0.5–10 mm/day) anterograde transport along microtubuli [Gamse et al., 1982]. Certain substances, like the nerve growth factor (NGF), are picked up at nerve terminals and forwarded by a retrograde transport to the cell body at 200 mm/day.

The Isolated and Cultured Dorsal Root Ganglion Cell

More recent studies were made in acutely isolated and cultured dorsal root ganglion neurons from the adult rat to establish the electrophysiological properties of afferent neurons. These cell bodies contain the neuropeptides substance P and CGRP, they respond to the algogen capsaicin and are sensitized in response to PGE_2. These cell bodies made it possible to study the electrophysiological properties in vitro [Gold et al., 1996]: application of bradykinin into isolated dorsal root ganglia caused a depolarization associated with an inward current and an increase in membrane conductance probably due to the opening of sodium channels. This process is mediated by a complicated activation of several enzymes [Burgess et al., 1989] and heightens the sensitivity of the Trpv receptor (see below).

Bradykinin

Bradykinin, a nonapeptide, is not produced by nerves and not released from nerves. It is, therefore, not a neurotransmitter. Bradykinin is split off from a plasma α-globulin by the enzyme kallikrein under a number of pathological conditions. It has many functions, of which the stimulation of sensory neurons is of interest.

When given by intra-arterial injection, it is the most active compound among many other algogenic substances. In addition, bradykinin releases PGE_2 from extraneuronal tissue for almost 1 h [Lembeck et al.,1976; Guo et al., 1999].

Nerve Growth Factor

Surprisingly, NGF shows effects similar to bradykinin. NGF is known as a survival factor for embryonic neurons and is essential for the development of all primary sensory neurons [Otten et al., 1980]. But in the adult, NGF serves a very different function; it is released from mast cells, fibroblasts and other cell types at sites of injury and inflammation, where it acts on primary sensory nerve terminals to promote thermal hypersensitivity. The exposure of cultured dorsal root ganglion neurons for several minutes produces acute sensitization to capsaicin-evoked responses. Although NGF elicits long-term changes in gene expression, its ability to promote short-term nociceptor sensitization suggests that posttranslational mechanisms are also involved [Chuang et al., 2001]. The enhanced activation of Trpv receptors is consistent with the potentiated effect of bradykinin. In summary, it has been found that rather complicated processes are involved in the stimulation and specifically the sensitization of nociceptors [Davis et al., 2000; for a review, see Julius and Basbaum, 2001; Holzer, 2004].

Capsaicin and Heat

Capsaicin is the pharmacological ingredient of paprika or chili *(Capsicum)*. It has a long history as a hot spice, but its pharmacological actions were discovered much later by Nicolas Jancsó, Aurelia Jancsó-Gábor and Janos Szolcsányi [1967] in Szeged.

In the first letter of Columbus to the Spanish Catholic Majesties Ferdinand and Isabella he already mentioned the habit of the people to eat meat with very hot spices. On his second voyage he was accompanied by the physician Doctor Diego Alvarez Chanca from Sevilla. He gave the first detailed report about the hot spice called Agi. In a letter in 1526 Cortez regarded Agi as one of the most valuable products of Mexico, saying that he enjoyed the hot taste of this spice. *Capsicum* was highly welcomed as a spice in Europe instead of the very expensive corns of black pepper (from *Piper nigrum* L.).

It is appetizing to see baskets full of green, yellow or red paprika fruits on the marked. These mild varieties of the fruits have a fine flavour but do not contain capsaicin. Only a fairly hot variety of *Capsicum*, called 'pepperoni', contains capsaicin [for a review, see Lembeck, 1987].

Small and Very Large Doses of Capsaicin

In very small doses (nanograms to micrograms) capsaicin immediately stimulates only Aδ/C fibres, as seen by vasodilatation and plasma protein extravasation, induced by the release of substance P and CGRP from peripheral terminals. After denervation, capsaicin is therefore inactive. Capsaicin makes it possible to analyze which actions are mediated by Aδ/C fibres.

On the other hand, treatment of newborn rats with a million-fold higher dose of 50 mg/kg induced a life-long degeneration of the Aδ/C fibres and a loss of substance P and other peptides present in these neurons. Such rats lose their sensitivity to heat and their heat loss reaction [Donnerer and Lembeck, 1983]. The rats no longer escape from a hot environment until they die from hyperthermia [Obal et al., 1979]. In capsaicin-pretreated rats the distension of the urinary bladder upon filling is no longer signalled to the CNS, while the efferent part of the reflex remains intact [Holzer-Petsche and Lembeck, 1984]. Otherwise the rats behave quite normally. But quite a number of reflex reactions and neuro-endocrine regulations were found to be inhibited or abolished [Lembeck and Donnerer, 2004].

Capsaicin-sensitive afferent fibres run to the spinal cord either with somatic nerves, like the sciatic nerve, or along with nerve fibres of the autonomic nervous system. The splanchnic nerve, usually regarded as a purely efferent nerve, contains 50% afferent neurons. Even the vagus nerve consists of about 90% afferent nerve fibres [for a review, see Lembeck and Donnerer, 2004].

Certain messages reach the level of consciousness like pain or changes in temperature. Other afferent fibres, which are 'switched off' by capsaicin, play a role in the induction of autonomic reflexes or as the link to neuro-endocrine regulation, not reaching the level of consciousness.

The Capsaicin Receptor

Recent years brought an essential new development. An expression cloning strategy was based on calcium influx to isolate a functional cDNA encoding a capsaicin receptor from sensory neurons. The receptor is called Trpv1 as it is sensitive to capsaicin (1 μmol/l) and to resiniferatoxin (0.1 μmol/l). Trpv1-expressing *Xenopus* oocytes were used to study electrophysiological responses [for a review, see Julius and Basbaum, 2001; Patapoutian et al., 2003]. The Trpv1 receptor is a non-selective cation channel and structurally related to the members of the transient receptor potential (TRP) channels. The capsaicin receptor is also activated by noxious heat; this suggests that it is the transducer of painful thermal stimuli in vivo [Caterina et al., 1997, 2000; Nagy and Rang, 1999], thus confirming earlier findings by Amann et al. [1990]. Jordt et al. [2000] demonstrated the stimulation of the capsaicin receptor by acid.

Birds are known to be insensitive to capsaicin. Their Trpv receptors are different but respond to noxious heat and acid [Jordt and Julius, 2002]. Mice depleted of the Trpv receptor show impaired nociception and reduced pain sensitivity.

The whole family of TRP channels might evolve as new interesting pharmacological targets [Gudermann and Flockerzi, 2005; Putney, 2004].

Menthol and Cold

Menthol is a naturally occurring compound which gives plants of the *Mentha* species their typical minty smell and flavour. The plant oil, usually referred to as peppermint oil (from *Mentha piperita*), is readily extracted from the plant by steam distillation. Before World War II it came from producers in Japan and China. When the trade routes were disrupted during World War II Brazil took over as the main producer. Peppermint oil is the world's third important flavouring agent behind vanilla and citrus flavours. It contains 30–50% menthol and the world production of peppermint oil is about 8,000 tons per year.

Menthol is used in a wide range of products, from common cold medications to tooth paste, from candy mints to chewing gum. Research indicated that menthol sensitizes and stimulates sensory cold receptors in the mouth and nose, and when applied externally also in the skin [for a review, see Eccles, 1994]. A synthetic compound, icilin, has a nearly 200-fold greater potency than menthol [Wei and Seid, 1983].

The menthol receptor, Trpm8 (also named CMR1, C for cold, M for menthol), has been characterized and cloned from trigeminal sensory neurons. It is a member of the TRP family of ion channels. The receptor was shown to be expressed by small diameter neurons in trigeminal and dorsal root ganglia. It was demonstrated that menthol and cold activate a calcium-permeable channel.

The receptor Trpm8 is expressed in a subpopulation of C (and possibly also Aδ) fibres representing less than 20% of all primary afferent fibres. Menthol and cold are detected by the same molecular entity on primary afferent neurons of the somatosensory system. The large proportion of nociceptive afferent fibres responsive to subzero temperatures indicate that, besides Trpm8, the heat-sensitizing receptors Trpv1 and Trpv2 are also activated in the noxious, pain-producing range of cold [for a review, see McKemy et al., 2002].

Overall, the C and Aδ primary afferents with all their different 'receptors' are able to respond to endogenous and exogenous noxious factors.

The functions of other major sensory systems such as olfaction, taste, vision and hearing are not mentioned here. They are all very complicated systems [for detailed information, please see the biochemistry textbook by Berg et al., 2002]. It should be mentioned that Richard Axel and Linda Buck received the 2004 Nobel Prize in Physiology or Medicine for their molecular biological research on olfactory receptors.

References

Amann R, Donnerer J, Lembeck F: Activation of primary afferents by thermal stimulation. Influence of ruthenium red. Naunyn Schmiedebergs Arch Pharmacol 1990;341:108–113.

Bayliss WM: On the origin from the spinal cord of the vasodilator fibres of the hind limb, and on the nature of these fibres. J Physiol (Lond) 1901;26:173–209.

Berg JM, Tymoczko JL, Stryer L: Biochemistry, ed 5. New York, Freeman, 2002.

Burgess GM, Mullaney I, McNeill M, Dunn PM, Rang HP: Second messengers involved in the mechanism of action of bradykinin in sensory neurons in culture. J Neurosci 1989;9:3314–3325.

Caterina MJ, Leffler A, Malmberg ABM, Martin WJ, Trafton J, Petersen-Zeitz KR, Koltzenburg M, Basbaum AI, Julius D: Impaired nociception and pain sensation in mice lacking the capsaicin receptor. Science 2000;288:306–313.

Caterina MJ, Schumacher MA, Tominaga M, Rosen TA, Levine JD, Julius D: The capsaicin receptor: a heat-activated ion channel in the pain pathway. Nature 1997;389:816–824.

Celander O, Folkow B: The nature and the distribution of afferent fibres provided with the axon reflex arrangement. Acta Physiol Scand 1953;29:359–370.

Chuang HI, Prescott ED, Kong H, Shields S, Jordt S-E, Basbaum AI, Chao MV, Julius D: Bradykinin and nerve growth factor release the capsaicin receptor from Ptdls(4,5)P_2-mediated inhibition. Nature 2001;411:957–962.

Dale HH: Pharmacology of nerve endings. Proc R Soc Lond Biol 1935;28:319–332.

Davis JB, Gray, Gunthorpe MJ, Hatcher JP, Davey PT, Harries MH, Latcham J, Clapham C, Atkinson K, Hughes SA, Rance K, Grau E, Harper AJ, Pugh PL, Rogers DC, Bingham S, Randall A, Sheardown SA: Vanilloid receptor-1 is essential for inflammatory thermal hyperalgesia. Nature 2000;405:183–187.

De Biasi S, Rustioni A: Glutamate and substance P coexist in primary afferent terminals in the superficial laminae of spinal cord. Proc Natl Acad Sci USA 1988;85:7820–7824.

Donnerer J, Amann R: Effect of NMDA receptor blockade on afferent nerve stimulation-induced autonomic reflexes. Naunyn Schmiedebergs Arch Pharmacol 1994;349:R85.

Donnerer J, Lembeck F: Heat loss reaction to capsaicin through a peripheral site of action. Br J Pharmacol 1983;79:719–723.

Eccles R: Menthol and related cooling compounds. J Pharm Pharmacol 1994;46:618–630.

Gamse R, Holzer P, Lembeck F: Decrease of substance P in primary afferent neurons and impairment of neurogenic plasma extravasation by capsaicin. Br J Pharmacol 1980;68:207–213.

Gamse R, Molnar A, Lembeck F: Substance P release from spinal cord slices by capsaicin. Life Sci 1979;25:629–636.

Gamse R, Petsche U, Lembeck F, Jancsó G: Capsaicin applied to peripheral nerve inhibits axoplasmic transport of substance P and somatostatin. Brain Res 1982;239:447–462.

Ganong WF: Medical Physiology. London, Appleton & Lange, 1991.

Gold MS, Dastmalchi S, Levine JD: Co-expression of nociceptor properties in dorsal root ganglion neurons from the adult rat in vitro. Neuroscience 1996;71:265–275.

Gudermann T, Flockerzi V: TRP channels as new pharmacological targets. Naunyn Schiedebergs Arch Pharmacol 2005;371:241–244.

Guo ZL, Symons JD, Longhurst JC: Activation and visceral afferents by bradykinin and ischemia: independent roles of PKC and prostaglandins. Am J Physiol 1999;276: H1884–H1891.

Hinsey JC, Gasser HS: The components of the dorsal root mediating vasodilatation and the Sherrington contracture. Am J Physiol 1930;92:679–689.

Holzer P: Capsaicin: cellular targets, mechanisms of action, and selectivity for thin sensory neurons. Pharmacol Rev 1991;43:143–201.

Holzer P: TRPV1 and the gut: from a tasty receptor for a painful vanilloid to a key player in hyperalgesia. Eur J Pharmacol 2004;500:231–241.

Holzer-Petsche U, Lembeck F: Systemic capsaicin treatment impairs the micturition reflex in the rat. Br J Pharmacol 1984;83:935–941.

Jancsó N, Jancsó-Gábor A, Szolcsányi J: Direct evidence for neurogenic inflammation and its prevention by denervation and by pretreatment with capsaicin. Br J Pharmacol 1967;31:138–151.

Jordt S-E, Julius D: Molecular basis for species-specific sensitivity to 'hot' chilli peppers. Cell 2002;108:421–430.

Jordt S-E, Tominaga M, Julius D: Acid potentiation of the capsaicin receptor determined by a key extracellular site. Proc Natl Acad Sci USA 2000;97:8134–8139.

Julius D, Basbaum A: Molecular mechanisms of nociception. Nature 2001;413:203–210.

Juránek I, Lembeck F: Evidence for the participation of glutamate in reflexes involving afferent, substance P-containing nerve fibres in the rat. Br J Pharmacol 1996;117: 71–78.

Juránek I, Lembeck F: Afferent C-fibres release substance P and glutamate. Can J Physiol Pharmacol 1997;75: 661–664.

Lembeck F: Zur Frage der zentralen Übertragung afferenter Impuse. III. Mitteilung: Das Vorkommen und die Bedeutung der Substanz P in den dorsalen Wurzeln des Rückenmarks. Arch Exp Pathol Pharmakol 1953;219: 197–213.

Lembeck F: Sir Thomas Lewis's nocifensor system, histamine and substance P-containing primary afferent nerves. Trends Neurosci 1983;6:106–108.

Lembeck F: Columbus, capsicum and capsaicin: past, present and future. Acta Physiol Hung 1987;69:265–273.

Lembeck F, Donnerer J, Tsuchiya M, Nagahisa A: The non-peptide antagonist, CP-96,345, is a potent inhibitor of neurogenic inflammation. Br J Pharmacol 1992;105: 527–530.

Lembeck F, Donnerer J: History of a pioneering neuropeptide: substance P; in Holzer P (ed): Tachykinins. Handb Exp Pharmacol. New York, Springer, 2004, vol 164, pp 1–23.

Lembeck F, Holzer P: Substance P as neurogenic mediator of antidromic vasodilatation and neurogenic plasma extravasation. Naunyn Schmiedebergs Arch Pharmacol 1979;310:175–183.

Lembeck F, Popper H, Juan H: Release of prostaglandins by bradykinin as an intrinsic mechanism of its algesic effect. Naunyn Schmiedebergs Arch Pharmacol 1976;294: 69–73.

Lembeck F, Sieler D: In vivo effects of the NMDA receptor agonist tetrazolyl-glycine related to the function of small-diameter primary afferents. Naunyn Schmiedebergs Arch Pharmacol 1999;360:38–41.

McKemy DD, Neuhausser WM, Julius D: Identification of a cold receptor reveals a general role for TRP channels in thermosensation. Nature 2002;416:51–58.

Nagy I, Rang H: Noxious heat activates all capsaicin-sensitive and also a sub-population of capsaicin-insensitive dorsal root ganglion neurons. Neuroscience 1999;88: 995–997.

Obal F, Benedek G, Jancso-Gabor A: Salivary cooling, escape reaction and heat pain in capsaicin-desensitized rats. Pflügers Arch 1979;382:249–254.

Otten U, Goedert M, Mayer N, Lembeck F: Requirement of nerve growth factor for development of substance P-containing sensory neurons. Nature 1980;287:158–159.

Patapoutian A, Peier AM, Story GM, Viswanath V: ThermoTRP channels and beyond: mechanisms of temperature sensation. Nat Rev Neurosci 2003;4:529–539.

Putney JW: Mammalian TRP Channels as Molecular Targets. Chichester, Wiley, 2004.

Stricker S: Untersuchung über die Gefässwurzeln des Ischiadicus. Sitzber Kaiser Akad Wiss (Wien) 1876;3:173.

Ueda M, Kuraishi Y, Sugimoto K, Satoh M: Evidence that glutamate is released from capsaicin sensitive primary afferent fibres in rats: study with on-line continuous monitoring of glutamate. Neurosci Res 1994;20:231–237.

Wei ET, Seid DA: AG-3-5: a chemical producing sensations of cold. J Pharm Pharmacol 1983;35:110–112.

Neurotransmission in the Central Nervous System

The Past

In the first edition of *The Pharmacological Basis of Therapeutics: a Textbook of Toxicology and Therapeutics for Physicians and Medical Students* [Goodman and Gilman, 1941], drugs acting on the central nervous system (CNS) were divided in two groups, depressants (general anaesthetics, hypnotics and sedatives, opiates and antipyretics) and stimulants (strychnine, metrazol and caffeine). Some of these have been used since prehistoric times (opium, belladonna and other plants), whereas others came from the New World (tobacco, cocaine). Others still stem from the early years of organic chemistry, such as anaesthetic gases, barbiturates and 'antipyretics'.

Quite a number of old remedies have been in therapeutic use for conditions or diseases which occur only in human subjects, such as insomnia, depression or mania, schizophrenia or epilepsy. It should not be forgotten that treatment and maintenance of patients with mental diseases was, until a few decades ago, nothing more than submitting them to a chain of cruelties.

The contrast between the first [Goodman and Gilman, 1941] and the ninth edition [Goodman et al., 1996] of *Goodman and Gilman's: The Pharmacological Basis of Therapeutics* is illustrated by the introduction to the chapter 'Neurotransmission and the Central Nervous System' by Floyd E. Bloom:

'Drugs that act upon the central nervous system (CNS) influence the lives of everyone, everyday. These agents are invaluable therapeutically because they can produce specific physiological and psychological effects. Drugs that affect the CNS may selectively relieve pain or fever, suppress disorders of movement, or prevent seizures. They may induce sleep or arousal, reduce the desire to eat, or allay the tendency to vomit. They may be used to treat anxiety, mania, depression or schizophrenia without altering consciousness.

The unique quality of drugs that affect the nervous system and behaviour places investigators who study the CNS in the midst of an extraordinary scientific challenge – the attempt to understand the cellular and molecular basis for the enormously complex and varied functions of the human brain.'

Many changes took place during these 56 years as some examples show: many anaesthetic gases disappeared, others remained and new ones were introduced. Former remedies, like bromide and antifebrin, were abandoned. A new use of aspirin was found, the prophylaxis of thrombosis. Self-administration became a widespread practice for certain hypnotics and anxiolytics.

Moreover, dramatic progress was made in the meantime. The number of neurotransmitters rose from about 10 'classical transmitters' to more than 50, including monoamines, peptides, amino acids, but also NO, eicosanoids, and nerve growth factors, acting in the periphery and within the CNS. The structures of their manifold receptors including their subtypes were investigated.

The pathways of classical neurotransmitters, acetylcholine, noradrenaline, dopamine and serotonin within the CNS were precisely mapped and partially attributed to certain functions (especially dopamine, see p. 154). It was difficult to define clearly the pathways of peptides; those of amino acid neurotransmitters act more universally.

The Neuron

Numerous new methods have made it possible to study in detail the biochemistry and functions of the various neurons in the CNS and only a few facts that are linked to neurotransmission are mentioned in the following.

A neuron consists of a cell body, the metabolic centre, with the nucleus carrying the genetic information. The dendrites pick up incoming signals with more than 10,000 synapses. The axon with its presynaptic terminal mediates the machinery of neurotransmission including the release, re-uptake and partly the metabolism of transmitters.

With the exception of the neurons of the hippocampus, the life span of the neurons in the CNS is as long as that of the individual.

Glia Cells

Rudolf von Virchow first described the neuroglia and regarded it as a sort of glue between neurones. The word 'glia' stems from the Greek word for glue. The CNS neurones are surrounded by a 9 times larger number of glia cells which provide essential functions:

(1) They are supporting elements, providing firmness and structure for the brain.

(2) Oligodendrocytes in the CNS are related to Schwann cells which isolate large-diameter axons by wrapping them in a tight spiral of myelin sheets, similar to a wire which is protected by an isolating tape. Schwann cells and macrophages around small-diameter sensory and sympathetic neurons produce the nerve growth factor which is taken up and transported to the cell body of the neuron in case of injury or a lesion.

(3) The number of microglia cells is comparable to the total number of neurons in the CNS. The microglia cells stem from early stages of haematopoietic cells. In situations of injury or apoptosis, as in oxygen-deficient regions, they change their shape and function. They serve as scavengers, removing the debris after injury or cell death. They are also activated in Alzheimer's disease, around brain tumours and in HIV encephalitis. HIV viruses in the brain are exclusively found in microglia cells.

(4) Astrocytes are the most numerous glia cells. Stephan Kuffler found that they have a high permeability for potassium ions [for details, see Kuffler et al., 1984]; they form a pericapillary membrane network which together with the endothelial cells and tight junctions form the blood-brain barrier. It limits and selects the entrance of substances into the brain [for details, see Kandel et al., 1991].

In recent years, more information about the function of glia cells has been obtained with the use of new methods, some of which might serve as examples: functional magnetic resonance imaging and positron emission tomography made it possible to localize activities with a millimetre resolution, but their time resolution was in the range of seconds or minutes. EEG has an excellent time resolution in the millisecond range, but the localization of different regions that contribute to the signals is a major problem. Therefore, some researchers used more than one analytical method.

These experiments explained the communication of neurons with glia cells. The membrane of glia cells contains potential-dependent ionic channels and receptors for neuronal messengers. It seems that glia cells 'listen' to signals from adjacent neurons and respond with the entrance of calcium ions. ATP was found to be involved in the response of glia cells which communicate to other neurons far away. This 'bridge function' might enable them to co-ordinate distant populations of neurons. In addition, astrocytes contain a vesicular compartment that is competent for regulated exocytosis of glutamate, somewhat similar to synapses [Bezzi et al., 2004].

Fields [2004] made a comparison: neurons can be regarded as telephones, mediating messages via cables to specific connections. Glia cells are like cell phones, sending messages all over. To receive the message, a specific receptor has to be switched on.

Astrocytes increase the number of mature functional synapses of the CNS and are necessary for synaptic maintenance in vitro. This raises the possibility that glia may actively participate in synaptic plasticity. The higher the stage of evolution of animals, the more glia in comparison to neurons is found. In mammalians and man the relative amount of glia and neurons is remarkably similar. The presence of glia cells in CNS tissue cultures increases the number of synapses formed by a factor of seven [Ullian et al., 2001]. Cholesterol has been found to be the factor to induce this multiplication of contacts between neurons. Neurons produce just enough cholesterol for survival, additionally needed cholesterol is produced by glia cells [Mauch et al., 2001].

It is known that the brain consumes about 20% of the metabolic energy of the body. Activation of hippocampus neurons increases their glycolytic metabolism for about 10 ms. Thereafter the neural activity triggers the glycolysis in astrocytes and the release of lactate by which the supply of the activated neurons is settled. This was recently shown by the two-photon fluorescence imaging in hippocampus slice preparations by the group of Webb [Kasischke et al., 2004]. This short description shows that glia cells cannot be regarded as silent servants. They have the ability to communicate by other means.

Assembly of the Adult Human Brain

Large anatomical divisions provide a classification of the distribution of brain functions. The adult human brain consists of about 100 billions of neurons connected to each another by still 1,000 times more synapses. The cortex finally covers all the other parts and amounts to 90% of the brain. It is the most complex organ known so far.

Cortex. The brain cortex consists of six layers which contain 100,000 neurons/mm^2. Columnar sets of about 100 vertically connected neurons are thought to form the elemental processing modules. Below the 3-mm-thick cortex are the nerve fibre connections to various regions.

The brain cortex is the core of all that makes us human, producing all higher cortical functions such as abstract thought and consciousness. It enables us to make plans or other considerations, or to understand a mathematical formula. It supervises the integration of the functions of the autonomic nervous system.

Diencephalon. The limbic system located closely to the diencephalon adds emotional and motivational functions and the hippocampus controls our memory (see below). The basal ganglia exert regulatory control of the voluntary motor system and regulatory control of visceral functions.

Midbrain and Brainstem. This part has a 'bridge function' between cortex and spinal cord and contains the nuclei of the cranial nerves and the reticular activating system.

Cerebellum. The cerebellum has a most detailed cytological organization. The motor orders of the cortex are under its constant control. Its continuing influence has a stabilizing function.

This rather superficial, and therefore incomplete, description of the assembly of the human brain becomes even more difficult, when it should be related to neurotransmission. The basic processes of transmission in the brain are essentially similar to those operating in the periphery. But all functions of the brain are basically operated by complicated networks. Neurotransmitters in the brain operate through receptors with fast and slow responses. The interneuronal connections in the brain are complex, and might be influenced by the release of amino acids from astrocytes. Adaptive responses like sensitization-desensitization might be involved. Therefore, the drug-induced interference with specific transmitter systems is sometimes difficult to explain.

Sleep

Everyone knows what sleep is, although it is difficult to define. There are two sleep patterns. The deep sleep which is interrupted by several REM periods. The activity of the cortex is reduced in deep sleep, and the firing rate of neighbouring neurons is synchronized as indicated by large slow waves in the EEG. This might help the recovery of the neurons from the effects of accumulated free radicals.

In the REM periods larger activity of the cortex and brain stem can be seen. It is the period of dreams. The release of the neurotransmitters noradrenaline, serotonin and histamine is depressed. This seems to allow the receptors to recover [Siegel, 2003].

The sleep under conditions of a modern society and the resulting consequences for health have recently been reviewed by Foster and Wulff [2005].

Memory

The distinction between the short-term memory, lasting for minutes, and the long-term memory which can last for days or years was first made by the American psychologist William James [1890]. The hippocampus is the only region in the brain where even in adulthood new nerve cells are formed, persisting for only 4 weeks. The hippocampus can be compared with a mail box of a rather limited storage capacity. To accept new messages, the existing ones have to be deleted. In the hippocampus, this means the continuous elimination and the formation of new neurons.

Selected or frequently received messages are transcribed to higher regions for permanent storage. For this, additional synapses have to be formed, which are provided by protein synthesis [Peters and Giese, 2003].

Fields [2005] reviewed the recent experimental progress. The short-term memory is based on a transient increase of the postsynaptic depolarization; this might increase the reuptake of neurotransmitters.

Strong and repeated stimuli of the synapse activate Ca^{2+} channels of the neuron. Activation of Ca^{2+} -dependent enzymes activates the transcription factor Creb (cyclic AMP-response element binding protein) which enters the nucleus. There Creb activates genes. The transcribed mRNAs leave the nucleus and translate into proteins which exclusively act on already 'potentiated' synapses. Thereby the transient memory is transformed into a permanent and irreversible memory.

The long-term memory can be recalled at a remarkable speed. A trained conference interpreter is able to translate from one language into another, even by changing the syntax, almost at real time. The brain of a trained tennis player is able to calculate simultaneously the flight coordinates of a tennis ball served at a speed of 100 mph and to adjust his precise motor reaction by continuously changing the position of his moving body.

The long-term memory has an enormous capacity. It can store hundreds of odours or still more melodies and instantly bring them back to our consciousness.

The induction of addiction depends of some sort of pleasant feelings after the ingestion of a drug which is stored, until the demand for a new dose becomes irresistible. With continuous use, the brain becomes dependent and adapts its functions to an altered normal level which breaks down upon immediate withdrawal of the drug.

The memory is also under the influence of the intellect. Otto Loewi expressed this quite simply: 'The memory is a function of interest.'

Treatment of Human Central Nervous Diseases

Diseases located in the brain have been known since ancient times. All kinds of treatments, without any rational basis, were tried in these patients. But the transfer into an insane hospital for the rest of their life was the final option. The number of beds needed for these patients amounted to about 50% of the total number of all clinical beds together.

For generations, four common diseases which occur only in man had been treated with a number of old remedies without success. It was, therefore, tempting to try new methods based on the advancing new knowledge in central neurotransmission.

Parkinson's Disease

Not long ago, the life of the millions of patients with Parkinson's disease was sad: stuck in a wheelchair, shaking and completely incapacitated, they finally died from poor nutrition, pneumonia and aspiration. The parasympathetic symptoms (salivation, sweating, incontinence) were the reason for giving them atropine in high doses, resulting in 20–30% improvement. This was the therapy for about 100 years. Today as a matter of routine levodopa is given as the 'gold standard' therapy. The substitution of the deficient release of the transmitter dopamine is the unique example of a substitution therapy in the brain. No other neurodegenerative disease can be treated equally well, though not cured (see p. 154).

Epilepsy

Epilepsy, characterized by seizures, results from episodic neuronal discharges. Various types of epilepsy (grand mal and petit mal) were defined and the therapy was adjusted accordingly. The EEG provides a well-defined diagnostic method. Animal models of the disease were helpful in the development of antiepileptic drugs. Their main mechanisms are (1) the enhancement of the action of GABA (see p. 172) by phenobarbital, benzodiazepines and a number of new drugs, and (2) inhibition of sodium and calcium channel function by phenytoin, carbamazepine, valproic acid and a number of newer drugs. The essential action of all the drugs is within the range of neurotransmission and is able to control the disease in about 70% of the patients.

Schizophrenia

Schizophrenia affects about 1% of the population and is, therefore, one of the most important forms of psychiatric diseases. The aetiology and pathogenesis include genetic and environmental factors. It is explained by neurochemical theories based on dopamine, glutamate and serotonin. Hardly any animal models exist which would be suitable for drug development. Antipsychotics take days to weeks to work, suggesting that secondary effects may be more important than direct effects on receptor blockade.

Depression

Depression is another common psychiatric condition. Animal models can hardly address the mood changes that define the condition in humans. Various types of the disease exist. There is good evidence that monoaminergic neurotransmission in the CNS is insufficient, and its pharmacological manipulation remains the most successful therapeutic approach.

Drugs Acting in the CNS

Opiates

Opium is the dried juice of *Papaver somniferum* which has been used since prehistoric times. Paracelsus carried opium-containing laudanum pills in the knob of his sword as his most important secret remedy. Sertürner [1803] isolated the main alkaloid of opium, morphine. Schaumann [1940] synthesized the first two synthetic opiate analgesics, pethidine and methadone, and introduced them to clinical use. A large number of most effective synthetic opiates have been discovered in the meantime.

In 1975, Hughes et al. [1975] succeeded in isolating from the brain two pentapeptides (enkephalins) that competed strongly with morphine-like drugs for binding to receptors in the brain. Their pharmacological actions closely resemble morphine (see pp. 197–200).

Morphine remains the only drug to act by mimicking these peptides on their specific receptors. The term opioid applies to any substance, whether endogenous or synthetic, that produces morphine-like effects which are blocked by its antagonist naloxone. Enkephalins are found in short interneurons of many brain areas, and β-endorphin in neurons that project from the hypothalamus to the thalamus.

The analgesic action of opiates is experimentally and clinically well established. But can opiates be called 'neurotransmitters' or better 'neuromodulators', when they act in the CNS and in the periphery not for seconds but for hours? They are in any case a separate entity.

Recently it has been found that everyone's brain contains endocannabinoids, marijuana-like substances. By a complicated mechanism, called depolarization-induced suppression of inhibition, they block the release of GABA. This process might be involved in pain, anxiety and vomiting [Nicoll and Alger, 2004].

General Anaesthetics

The anaesthetic gases nitrous oxide, ethyl ether and chloroform were introduced between 1842 and 1850. They started to be used without previous animal experiments directly on human patients. The dramatic history of these discoveries is found in earlier textbooks such as *Goodman and Gilman's Textbook of Pharmacology* [1941]. Surgical anaesthesia became possible because the exclusion of cortical functions and of spinal motor reflexes commenced earlier, while the brain stem with the vital respiratory and circulation reflex centres remained functioning. The anaesthetic gases are not metabolized and leave the body unchanged. Meyer [1937] wrote: 'Narcosis commences when any chemically indifferent substance has attained a certain molar concentration in the lipids of the cell.' The 'lipid theory' by Meyer (Vienna) and Overton [1901] (Zürich), in 1898, indicated that the anaesthetic potency of these gases corresponds with their lipid/water distribution coefficient. This made it possible to conclude that the lipids of the membrane are their primary target, not excluding lipid-protein interfaces of the membrane. These proteins seem to inhibit reversibly excitatory receptors (glutamate, acetylcholine, serotonin) and facilitate the function of those which are inhibitory (GABA, glycine). Franks and Lieb [1994] showed that general anaesthetics at pharmacologically relevant concentrations affect ligand-gated ion channels. The role of second messengers remains uncertain and anaesthetics act directly on proteins rather than on lipids. Tao and Johns [2004] described recently a group of postsynaptic proteins called the postsynaptic density zone that might be a new molecular target for inhalational anaesthetics.

'Antipyretics'

After their introduction, the group of non-steroidal anti-inflammatory drugs were widely used in the clinic. At around 1880, 90% of the hospitalized patients had a fever. Its reduction, first by antifebrin and then by several other drugs, seemed to be helpful.

Nowadays most of them are used by self-administration. Antipyrine and analogues, like diclofenac or iboprufen, mainly serve the inhibition of inflammatory pain, while a new indication was found for aspirin, that is the prophylaxis of thrombosis. The essential action is the inhibition of the cyclo-oxygenase and, therefore, the production of prostaglandins and thromboxanes.

The inflammatory reaction is inhibited in the periphery. Their analgesic effect is assumed to be based on the reduction of inflammation, but also includes a central analgesic action. The antipyretic action is assumed to be based on the inhibition of the prostaglandin production in the hypothalamus. There is no connection to an influence on central neurotransmission.

Barbiturates
Fischer and von Mering [1903] introduced diethylbarbiturate which eliminated the use of chloral hydrate, synthesized by Liebig in 1832, and used for insomnia since 1869. Hundreds of derivatives of barbiturates followed; the lipid-soluble thiobarbiturates for the first time made surgical anaesthesia by the intravenous route possible. Barbiturates seem to inhibit the CNS in a non-selective manner, in the range between sedation, reduction of anxiety and induction of sleep, including unconsciousness; respiratory depression and cardiovascular failure signal a toxic level of the drug. In the meantime they have been substituted by benzodiazepines.

Meprobamate
Berger and Bradley [1946] incidentally found mephenesin that reduced the facilitation of polysynaptic reflexes by the action on spinal interneurons. In the following pharmacological screening meprobamate was discovered as the most active compound inducing muscle relaxation. In the practical application, however, the simultaneous loss of anxiety dominated and it soon found a widespread use under the name of 'Miltown', the first drug called a 'tranquilizer'.

Chlorpromazine
Another discovery was made by Laborit and Huguenard [1951, 1952]. They attempted to produce in patients what they called 'artificial hibernation'. Derived from antihistamines of the phenothiazine type, chlorpromazine caused a deep central depression which allowed to lower the body temperature artificially; this seemed to facilitate the performance of long-lasting operations [Steinbereithner et al., 1955].

The antipsychotic potential of chlorpromazine was demonstrated by Delay and Deniker [1953]. It was found to be unique in controlling the symptoms of psychotic patients without excessively sedating them. Pharmacological screening showed that it blocked the action of many transmitters and this multiplicity led to the trade name Largactil. It is now clear that antagonism at dopamine receptors is the main determinant of the antipsychotic action.

Benzodiazepines
At the same time pharmaceutical companies, stimulated by the success of the previously mentioned 'tranquilizers' and 'antipsychotics', conducted large neuropharmacological screening programmes in laboratory animals. An essential new approach started with the synthesis of compounds with the unusual 7-ring, introduced by Sternberg. Such a compound was included in the screening program of Hoffmann-La Roche. The compound was without any effect in all these tests, but the technicians in the pharmacological laboratory insisted on their observation that the rats were peculiarly 'tame' when handled. In a test in the Basel Zoo on recently arrived, anxious or aggressive animals, the simple observation in the laboratory was confirmed. This was the discovery of chlordiazepoxide. Hundreds of further benzodiazepines have been synthesized and investigated since then.

The mode of action of benzodiazepines is based on the postulate that its receptor exists in two conformations, only one of which can bind to a molecule of the central inhibitor GABA. Normally there is an equilibrium between the two conformations, but the sensitivity to GABA is substantial. Benzodiazepines, postulated to bind preferentially to conformation A, shifts the equilibrium in favour of A and enhances the sensitivity to GABA. The resulting main actions of benzodiazepines are (1) reduction of anxiety and aggression, (2) sedation and induction of sleep, (3) reduction of muscle tone and co-ordination, (4) an anticonvulsive effect and (5) anti- and retrograde amnesia. Benzodiazepines indicate that their influence on a neurotransmitter process within the CNS exists, but is much more complex that action at peripheral synapses.

CNS Stimulants

Psychomotor stimulants like amphetamines release catecholamines and retard their uptake. Cocaine and mescaline have similar actions. Methylxanthines, like caffeine, inhibit phosphodiesterase and thus increase intracellular cAMP. Psychotomimetics like LSD and psilocybin act on serotonin receptors. Their hallucinogenic activity is, however, difficult to ascribe to a transmitter mechanism. Strychnine, a convulsant poison of no therapeutic value, blocks the action of released glycine, which inhibits interneurons in the spinal cord. It is the only alkaloid which interferes with the action of an amino acid.

Development of Intellectual Functions

All our receiving and expressing intellectual functions are based in the cerebral cortex. The cortex of the human brain is not only outstanding in its size of about 90% of the total brain, covering all the other parts.

Its development had been regarded as a time-lapse record of the human evolution. Recent findings, however, indicate that its development is more complicated, citing Shreeve [2005]:

'Four weeks after conception, the embryo was producing half a million neurons every minute. Over the next several weeks these cells migrate to the brain, to specific destinations by genetic cues and interactions with neighbouring neurons. During the first and second trimesters of the pregnancy, the neurons began to reach tentacles out to each other, establishing synapses, at a rate of two millions a second. Three month before birth the brain possesses more cells than needed as a fetus and even far more than needed as an adult. Then, just weeks away from birth, the trend reversed. Groups of neurons competed with each other to recruit other neurons into expanding circuits with specific functions. Those that lost died off in a pruning process scientist call "neural Darwinism". At birth, the predisposition to the mother's voice already exists. For the next 18 months, the baby's brain is already a learning machine. The last of her sense to develop fully is the vision. Even so, the baby clearly recognizes the mother's face at just two days old.'

Neuronal connections are discarded and others are enforced. New functions develop. At 6 months, the baby forms his first words. At one and a half years, he knows how to use his first 50 words. Their number thereafter seems to explode; in the 3rd year sentences are formed according to the rules of grammar. The memory starts to function. When puberty is reached, the brain is near to a linear structure, but open to further adaptations throughout life.

Paleontologists have helped us in recent years to understand the early evolution of man. More than 6 million years ago the evolution of man started with his separation from African apes. Development followed in steps over a long time period. When finally *Homo sapiens* arrived in Europe, 150,000–50,000 years ago, he brought with him essential new abilities. With his hands he had the skill to make all kinds of tools. Human figures made from clay and flutes from bones might reflect the level of his social life. Burial addition made from stones or ivory were not merely decorative and expressed that he had developed ideas about a continuity of life after death. He was able to remember the three-dimensional impression of a flying animal and to draw a two-dimensional coloured picture of the animal onto the wall of a cave. And most important, he also brought with him the ability to communicate in sentences.

This highly developed cortex consists of an uncountable number of neurons whose cell bodies are 10 µm in size with more than 10,000 synaptic contacts on their dendrites. This small size of the smallest 'units' allows extremely fast functioning.

Language – Guideline for the Intellect

Language was the first means to transmit precise and detailed information. Much later, at historical times, language was the bridge to writing. Symbols like the Egyptian hieroglyphs, the Chinese signs and our many modern pictograms were later expressed by 25 letters. With five horizontal lines and a small number of symbols, a melody can be documented. By writing and reading information can be stored forever. How many generations ago did Ovid write his *Metamorphoses* or Pythagoras formulate his $a^2 + b^2 = c^2$? Only man can store his knowledge over generations. As Medawar and Medawar [1977] wrote:

'Only human beings guide their behaviour by a knowledge of what happened before they were born and a preconception of what may happen after they are dead; thus only human beings find their way by a light that illuminates more than the patch of ground they stand on.'

All higher mental functions are based in our brain and can voluntarily and at once be called to our consciousness. There is so much we see, hear or read throughout our life. But this does not mean that we can explain or understand all this, such as the imagination of a composer or a painter, the constructive ideas of an architect or a philosopher, the conducting of life according to the command of a religious dogma, to mention only a few examples.

Neuroscience has been growing fast during the recent decades, as simply illustrated by the size of the American Society for Neurosciences, from 500 members in 1970 to 28,000 members today. The strategy was essentially 'bottom-up', going into more and more details of structure and function of neurons and glia, facilitated by molecular biology, immunological methods and genetics. On the other hand, a 'top-down' strategy, the research of cortical functions, followed more slowly.

Let us try to show this with an example. We know all the details of the material (wood, brass, strings) and also the construction of all instruments used in an orchestra. We can analyze the physics of the produced sounds and can store them by electronic devices for exact reproduction. We know about the precise coordination of a hundred players by the conductor. But the essential final effect of the concert, 'what we feel in our heart', what induces enjoyment or tristesse in our mind and finally 'moves us', remains beyond an approach by brain research.

Our brain research is 'technically' oriented. We can be proud of what we have reached so far. Medical success in diagnosis and therapeutics was achieved. Based on evidence, confirmed by a thousandfold use, an advance in therapeutics has been made.

In a lecture about 'The Borders of Our Knowledge in Natural Sciences' the physiologist Emil du Bois-Reymond [1872] defined these borders and mentioned 'the incomprehensible', the mind. According to our knowledge based on materialistic facts, its function cannot be explained. He concluded finally: 'Ignoramus – ignorabimus' (we do not know and we will never know). Even today, with so much additional information, we still have to confess the 'ignorabimus', which Otto Loewi expressed in the following simple words:

'The beauty of a quartet can never be explained by some acetylcholine in nerves or muscles or by any other chemical reaction.'

References

Berger FM, Bradley W: The pharmacological properties of myanserin. Br J Pharmacol 1946;1:265–272.

Bezzi P, Gundersen V, Galbete JL, Seifert G, Steinhäuser C, Pilati E, Volterra A: Astrocytes contain a vesicular compartment that is competent for regulated exocytosis of glutamate. Nat Neurosci 2004;7:613–620.

Delay J, Deniker P: Neuroplegics in psychiatric therapy. Therapie 1953;8:347–364.

Du Bois-Reymond E: 1872; quoted from Singer W: Frankfurter Allgemeine Zeitung, 2000, No 222, p 52.

Fields RD: The other half of the brain. Sci Am 2004;290: 54–61.

Fields RD: Making memories stick. Sci Am 2005;292(2):75–81.

Fischer E, von Mering J: Über eine neue Klasse von Schlafmitteln. Ther Gegenw 1903;44:97–101.
Foster RG, Wulff K: The rhythm of rest and excess. Nat Rev Neurosci 2005;6:407–414.
Franks NP, Lieb WR: Molecular and cellular mechanisms of general anaesthesia. Nature 1994;367:607–614.
Goodman L, Gilman A: The Pharmacological Basis of Therapeutics: a Textbook of Toxicology and Therapeutics for Physicians and Medical Students, ed 1. New York, Macmillan, 1941.
Goodman LS, Limbird LE, Milinoff PB, Ruddon RW, Goodman Gilman A: Goodman and Gilman's: The Pharmacological Basis of Therapeutics, ed 9. New York, McGraw-Hill, 1996.
Hughes J, Smith TW, Kosterlitz H, Fothergrill LA, Morgan BA, Morris HR: Identification of two related pentapeptides from the brain with potent opiate agonist activity. Nature 1975;258:577–579.
James W: The Principles of Psychology. New York, Holt, 1890.
Kandel ER, Schwartz JH, Jessell TM: Principles of Neural Science, ed 3. Norwalk, Appleton & Lange, 1991.
Kasischke KA, Vishwasrao HD, Fisher PJ, Zipfel WR, Webb WW: Neural activity triggers neuronal oxidative metabolism followed by astrocyte glycolysis. Science 2004;305: 99–103.
Kuffler SW, Nicholls JG, Martin AR: From Neuron to Brain: a Cellular Approach to the Function of the Nervous System, ed 2. Sunderland, Sinauer, 1984.
Laborit H, Huguenard P: Artificial hibernation by pharmacodynamical and physical means. Presse Méd 1951;59: 1329.
Laborit H, Huguenard P: Present technic of artificial hibernation. Presse Méd 1952;60:1455–1456.
Mauch DH, Nägler K, Schumacher S, Göritz C, Müller EC, Otto A, Pfrieger FW: CNS synaptogenesis promoted by glia-derived cholesterol. Science 2001;294:1354–1357.
Medawar PB, Medawar JS: The Life Science: Current Ideas of Biology. New York, Harper and Row, 1977.
Meyer KH: Contributions to the theory of narcosis. Faraday Soc Trans 1937;33:1062–1064.
Nicoll RA, Alger BE: The brain's own marijuana. Sci Am 2004;291(6):68–75.
Overton E: Studien über die Narkose. Jena, Verlag Gustav Fischer, 1901.
Peters M, Giese KP: Discovering novel subtypes of long-term memory. BIF Futura 2003;18:231–239.
Schaumann O: Arch Exp Pathol Pharmakol 1940;196:109.
Sertürner F: J Pharm Ärzte Apotheker 1805;13:229–243.
Shreeve J: Beyond the brain. Natl Geogr 2005;207:2–31.
Siegel JM: Why we sleep. Sci Am 2003;289:92–97.
Steinbereithner K, Lembeck F, Hift ST: 'Künstlicher Winterschlaf'. Munich, Urban & Schwarzenberg, 1955.
Tao YX, Johns RA: Neuronal PDZ domains. Mol Intervent 2004;4:215–221.
Ullian EM, Sapperstein SK, Christopherson KS, Barres BA: Control of synapse number by glia. Science 2001;291: 657–661.

From Transmitter to Receptor: Progress within 50 Years

Since Alquist [1948] introduced the concept that there are distinct types of cell membrane receptors for noradrenaline, mediating separate functions, pharmacologists have attempted to characterize the receptors for many neurotransmitters and hormones. Rigorous attention to receptor characterization and classification of selective synthetic ligands has led to major therapeutic advances, as exemplified best by the introduction of β-adrenoceptor- and H_2 histamine receptor-blocking drugs [Black et al., 1972; Black and Prichard, 1973]. Nevertheless, until the advent of modern molecular biology techniques, pharmacologists were sceptical about data from functional studies in whole tissues that suggested a multiplicity of receptor types for a given mediator.

Receptor gene cDNAs are now being cloned so rapidly that new receptor types are being proposed before their function in intact tissue, if any, is known. This creates problems for those involved in the classification and nomenclature of receptors potentially relevant to therapeutics.

A strong international desire to rationalize the information provided by techniques of molecular biology in relation to drug-related data concerning receptor function led to the activity of the International Union of Pharmacology (IUPHAR).

The IUPHAR set up the Committee for Receptor Nomenclature and Drug Classification. To provide the necessary scientific expertise, the committee has branched out into 19 subcommittees. The prepared documents are finally approved by the Executive Committee of the IUPHAR and submitted for publication in *Pharmacological Reviews* [Vanhoutte et al., 1994].

At the IUPHAR Congress 2002 in San Francisco, Calif., it was stated that *Pharmacological Reviews* are a most effective vehicle for the wide dissemination and adoption of the IUPHAR nomenclature recommendations. The *Pharmacological Reviews* had an impact factor of 25, the highest of all pharmacological publications, to which the IUPHAR nomenclature reports contribute [Abernethy and Ruffolo, 2002].

A 'Guide to Receptors and Channels' by Alexander et al. [2004] has recently been published.

The impact of the recent development in teaching pharmacology can be illustrated by the fact that nearly 6% or 45 pages of the text is devoted to molecular aspects of receptors, as seen in the 5th edition of Rang et al. [2003] under the titles 'How Drugs Act: Molecular Aspects' and 'Cellular Mechanisms: Excitation, Contraction, Secretion'.

References

Abernethy DR, Ruffolo RR: International Union of Pharmacology Nomenclature and Pharmacological Reviews. Pharmacol Rev 2002;54:159.

Alexander SPH, Mathie A, Peters JA: Guide to receptors and channels. Br J Pharmacol 2004;141(suppl 1).

Alquist RP: A study of the adrenotropic receptors. Am J Physiol 1948;153:585.

Black JW, Duncan WAM, Durant CJ, Ganellin CR, Parsons EM: Definition and antagonism of histamine H2-receptors. Nature 1972;236:385–390.

Black JW, Prichard BNC: Activation and blockade of β adrenoceptors in common cardiac disorders. Br Med Bull 1973;29:163–167.

Rang HP, Dale MM, Ritter JM, Moore PK: Pharmacology, ed 5. Edinburgh, Churchill Livingstone, 2003.

Vanhoutte PM, Barnard EA, Cosmides GJ, Humphrey PA, Spedding M, Godfraind T: International Union of Pharmacology Committee on Receptor Nomenclature and Drug Classification. Pharmacol Rev 1994;46:111–116.

SUBJECT INDEX